High Bridge, Kentucky

Melissa C. Jurgensen

Second edition 2019

Melissa C. Jurgensen

ISBN-10: 0615780776

ISBN-13: 978-0615780771

Cover photo: High Bridge - Melissa C. Jurgensen

This book is dedicated to the memories of Clyde Bunch and Ken Houp.

Titles by Melissa C. Jurgensen

High Bridge, Kentucky

Through Their Eyes: Covered Bridges of Bourbon County, Kentucky

Through Their Eyes: Covered Bridges of Harrison County, Kentucky

Through Their Eyes: Covered Bridges of Fleming County, Kentucky

Through Their Eyes: Covered Bridges of Franklin County, Kentucky

River Towns of Central Kentucky

Kentucky's Covered Bridges (with Laughlin)

Contents:

High Bridge

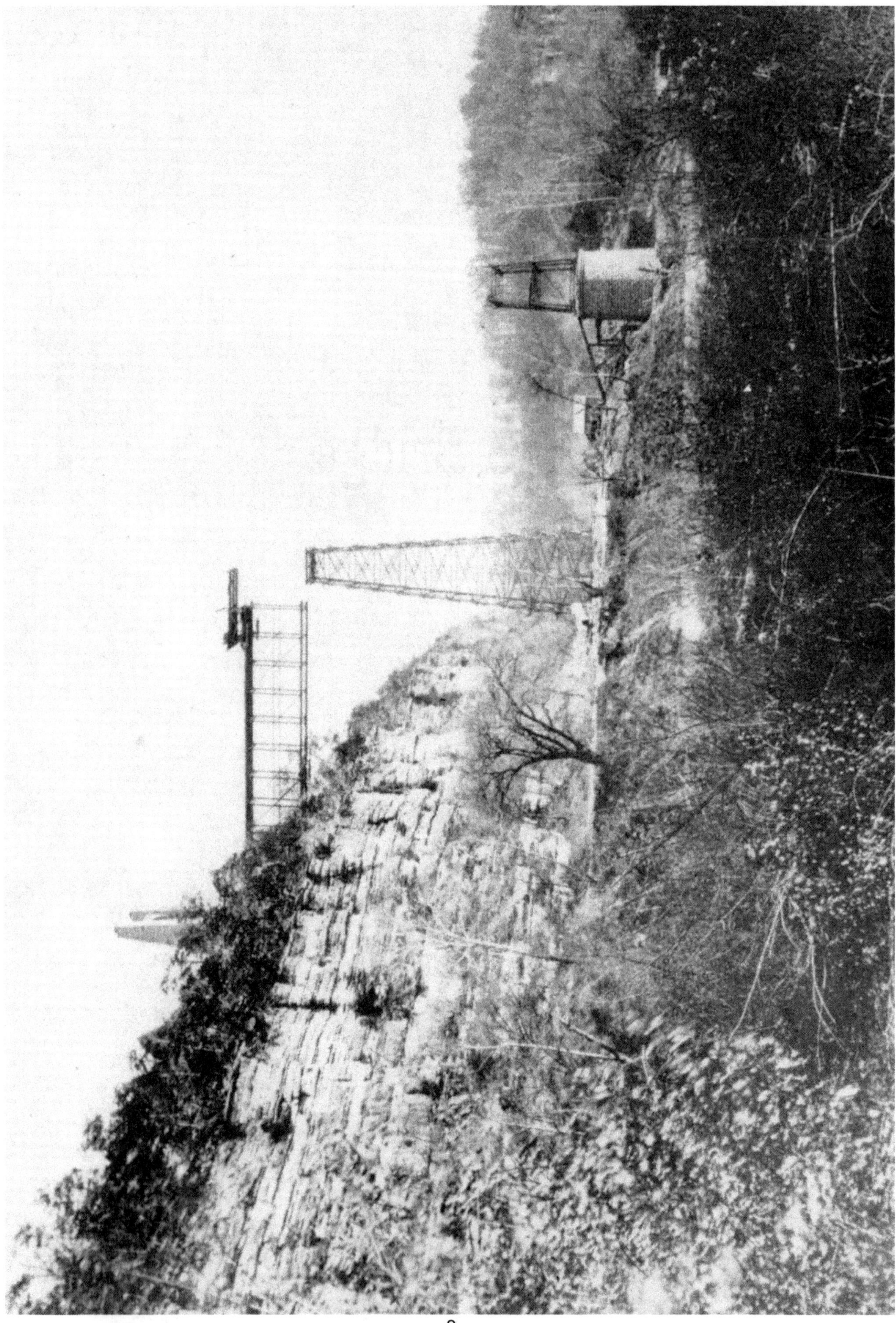

In the years before the Civil War, there was a great desire for a railroad to directly connect central Kentucky to Cincinnati, Ohio, to the north and to cities in the southern United States without having to send freight to these destinations through Louisville. The Lexington and Danville Railroad Company was formed to undertake the project between those two towns, but there were two major obstacles: the Kentucky River and the limestone cliffs that ran along side it. Jonathan A. Roebling, who would engineer New York City's Brooklyn Bridge later in his life, was contracted to build a suspension railroad bridge across the river gorge at the Jessamine and Mercer County lines. Roebling began construction of the stone towers for the suspension cables in 1858, but due to the escalation of the Civil War, his massive suspension bridge was never built. His towers stood alone on the cliffs, never to be used.

Planning for a bridge resumed after the Civil War. The Cincinnati Southern Railroad was chartered in 1874 and bought out the remaining shares of the Lexington and Danville Railroad Company. Cincinnati engineer Charles Shaler Smith was engaged to design a bridge across the river. Rather than utilizing the Roebling towers, he designed a bridge to be constructed of three 375-foot spans that would rest on piers with rock footings. The building of his engineering masterpiece began in October 1876, and was completed in February 1877.

During the bridge's construction, people traveled to High Bridge, then known as North Towers, to watch the work progress. When it was open, the bridge and the area around it became a popular tourist attraction. The bridge carried trains across the Kentucky River up into the Twentieth Century.

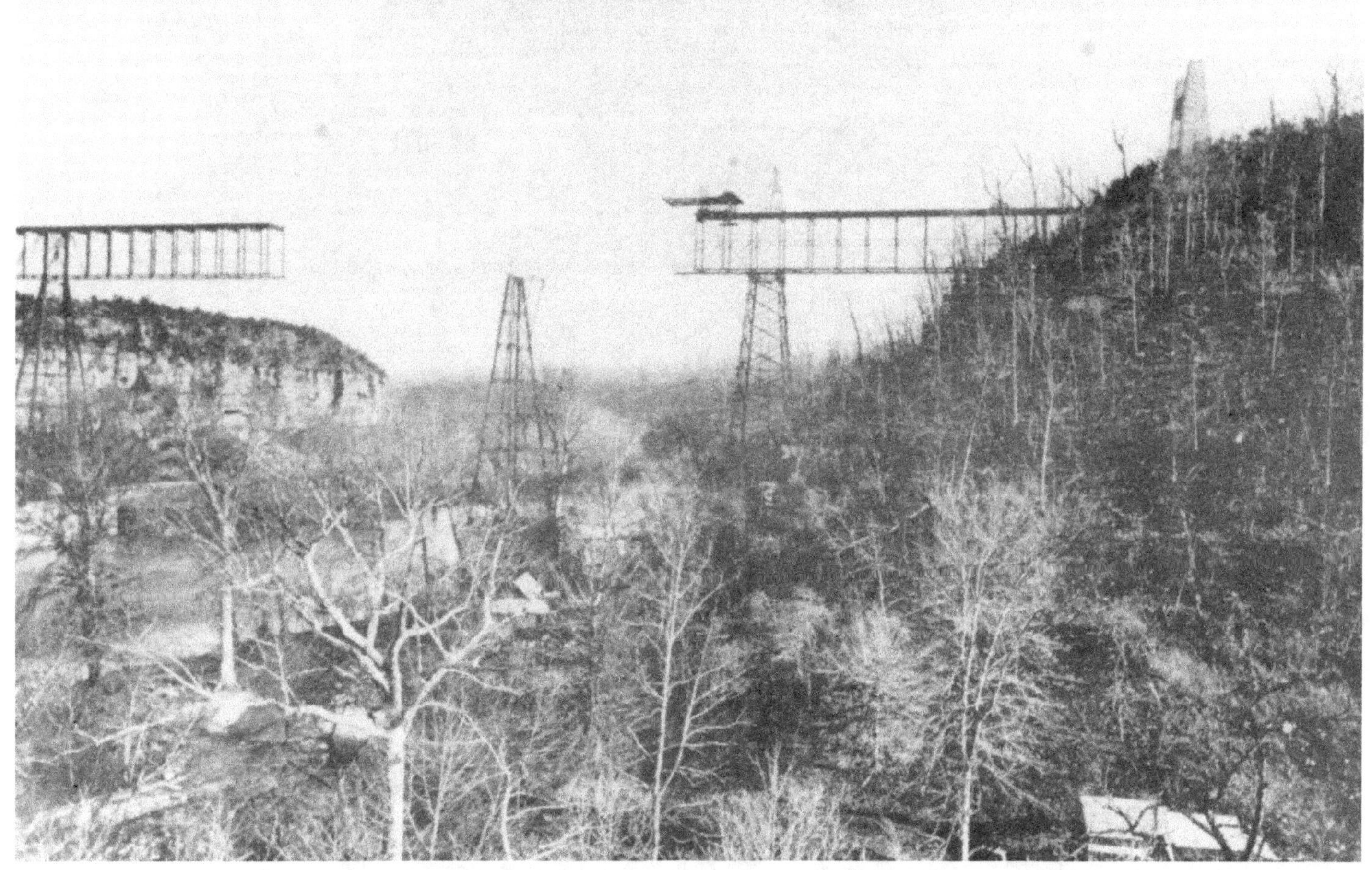

Rather than building the bridge from one side of the gorge to the other, work began on each side of the gorge simultaneously and the spans met in the middle. This January 23, 1877, image shows the work progressing toward a temporary support pier in the center of the bridge.

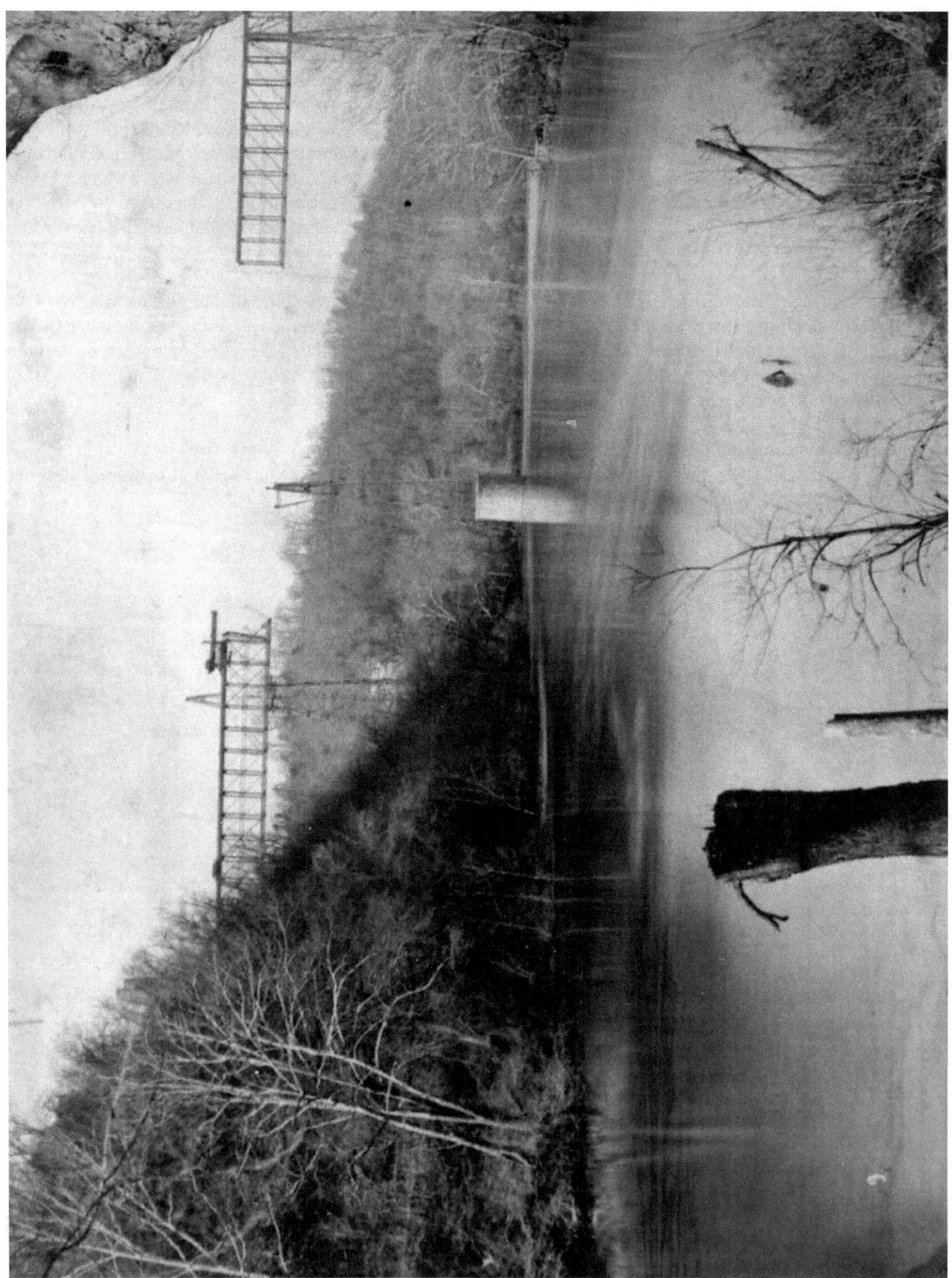

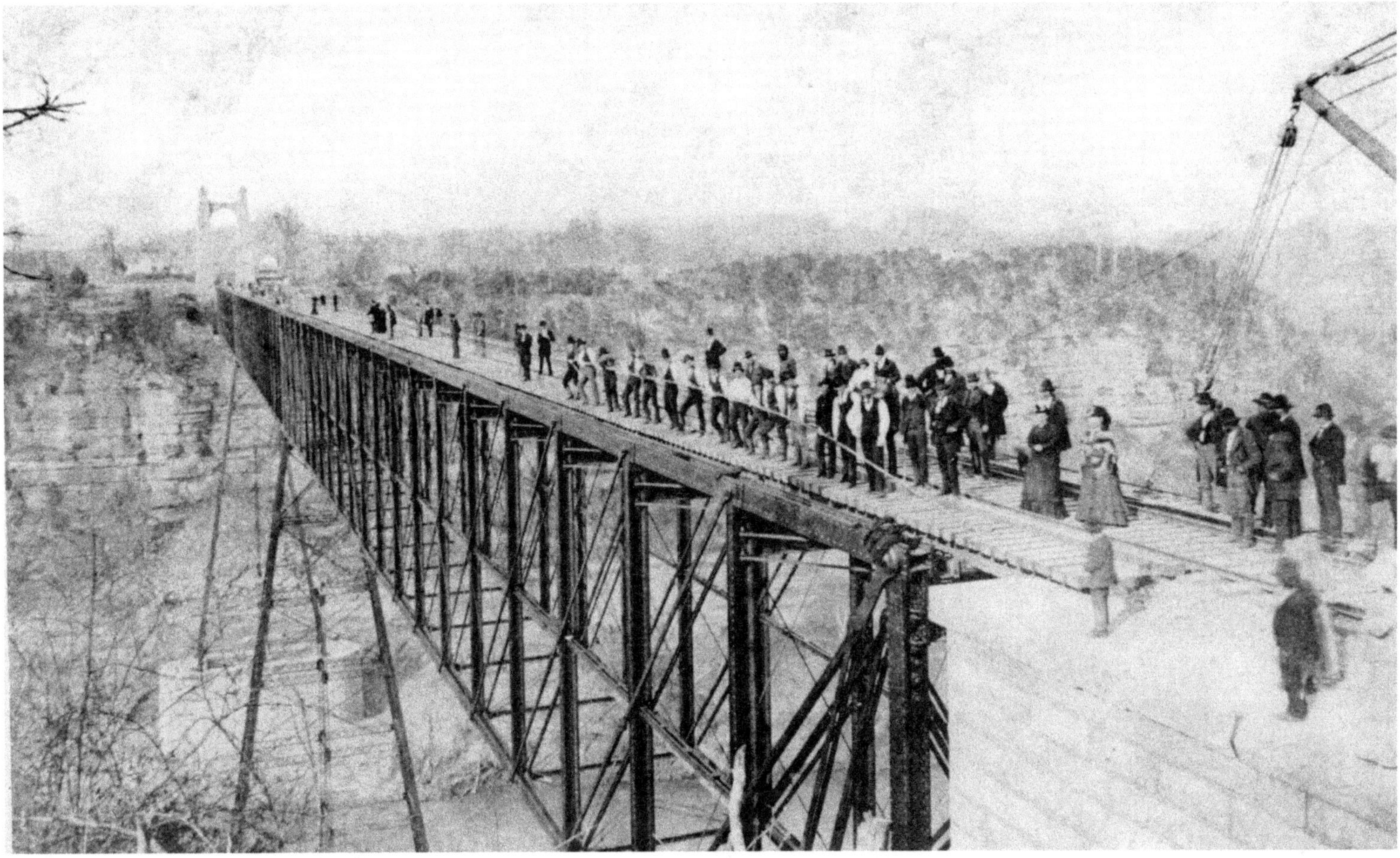

Though there were many men working on the bridge, and conditions could be quite dangerous, that did not stop onlookers from traveling to the bridge to watch the progress of the work. This helped establish High Bridge as a gathering place in the hearts and minds of Kentuckians before the bridge was even completed.

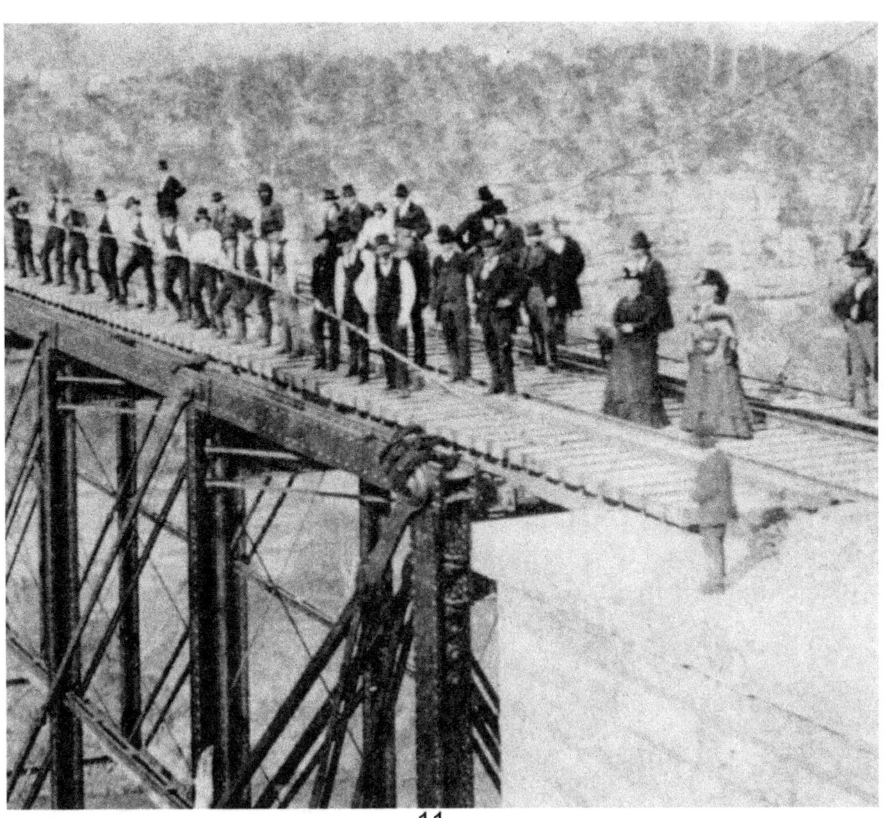

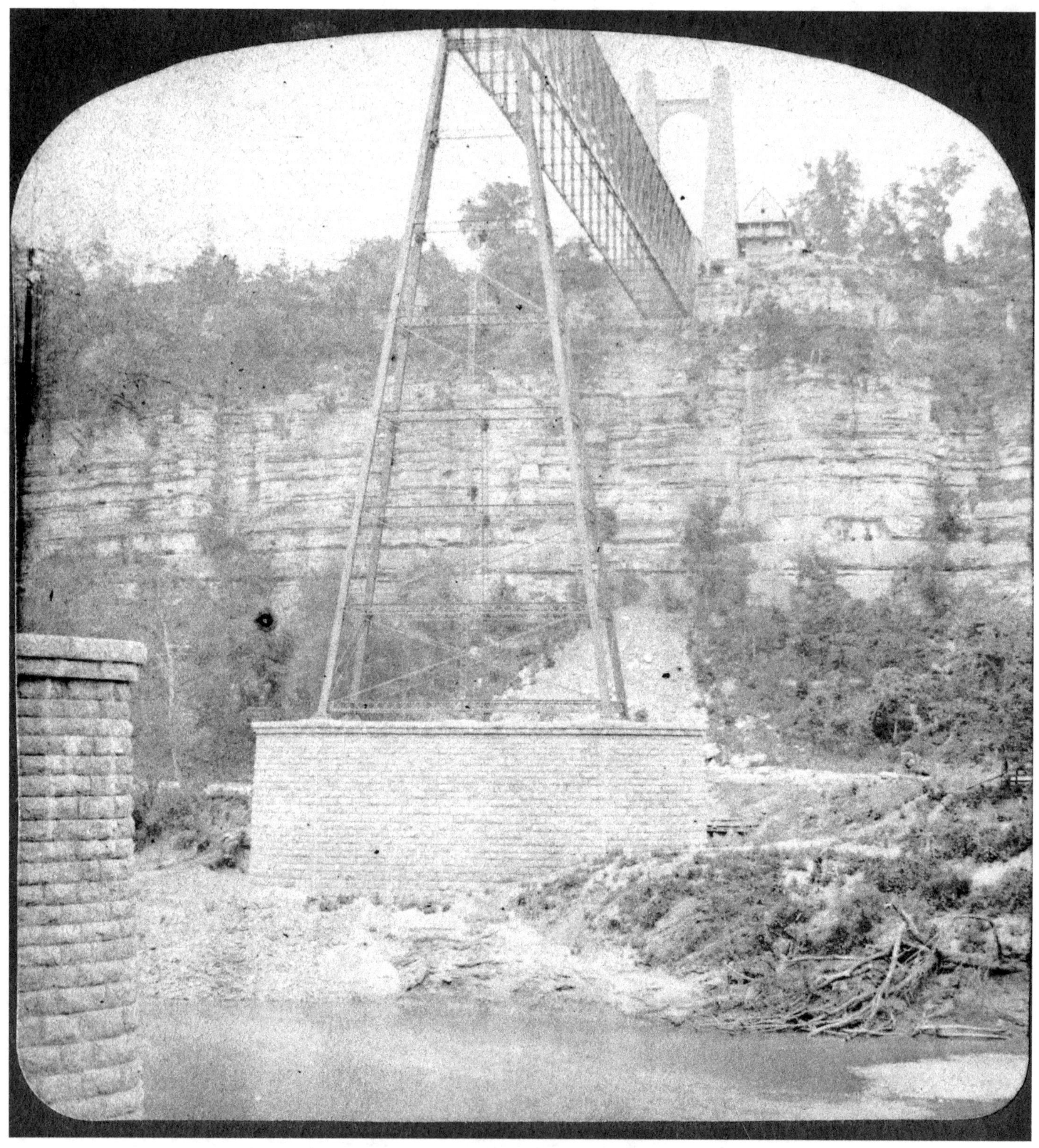

Local lore has it that the bridge workers could not get the last piece of iron to go into place because it was four inches too long. When a call was placed to Charles Shaler Smith, the designer in Cincinnati, he told them to wait until nightfall because, he claimed, "if it gets as cold as I calculated it would be, it will fit perfectly." Allegedly, it did.

The completed bridge stood 286.1 feet above the pier base, 275 feet above the low water mark, and was 1,125 feet long. The total cost for the project is estimated between $403,000.00 and $411,000.00. Until the early Twentieth Century, it was the highest bridge over a navigable stream in the world.

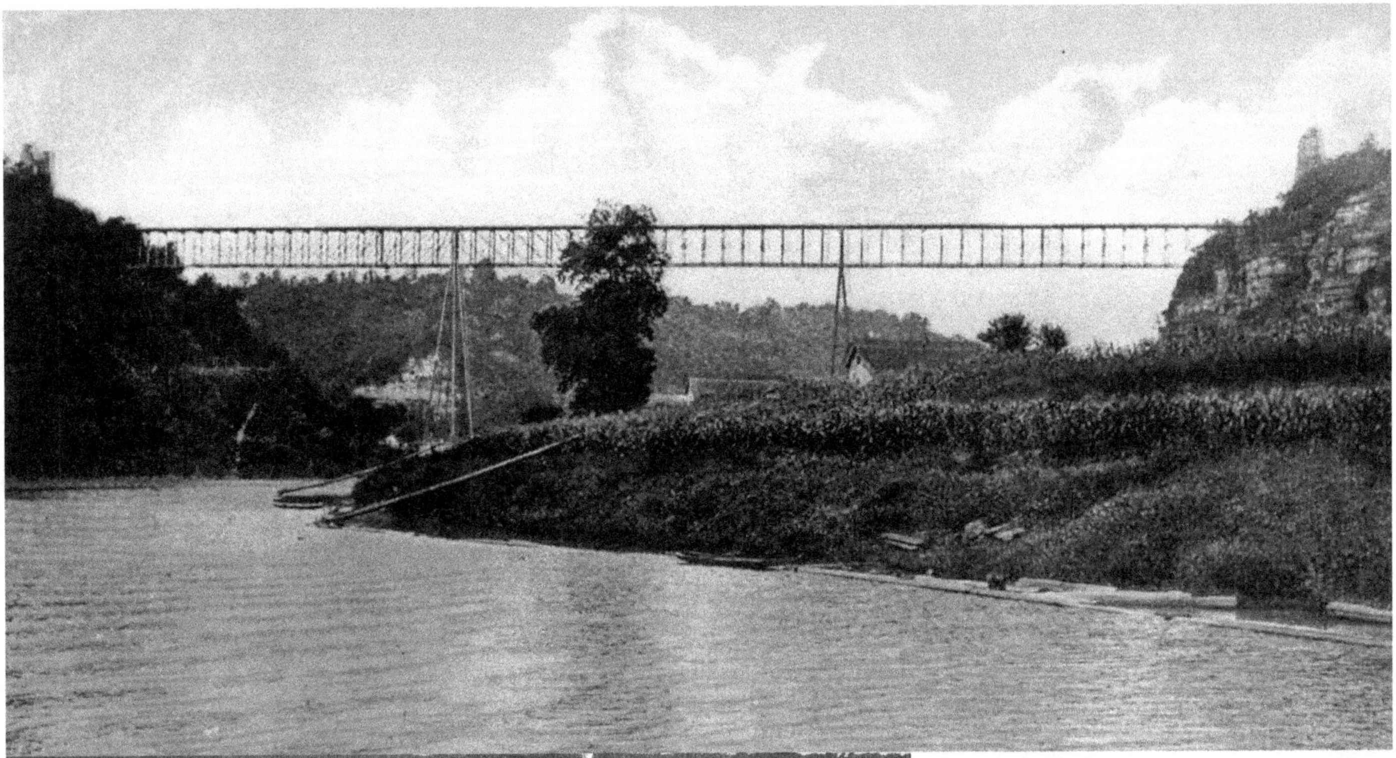

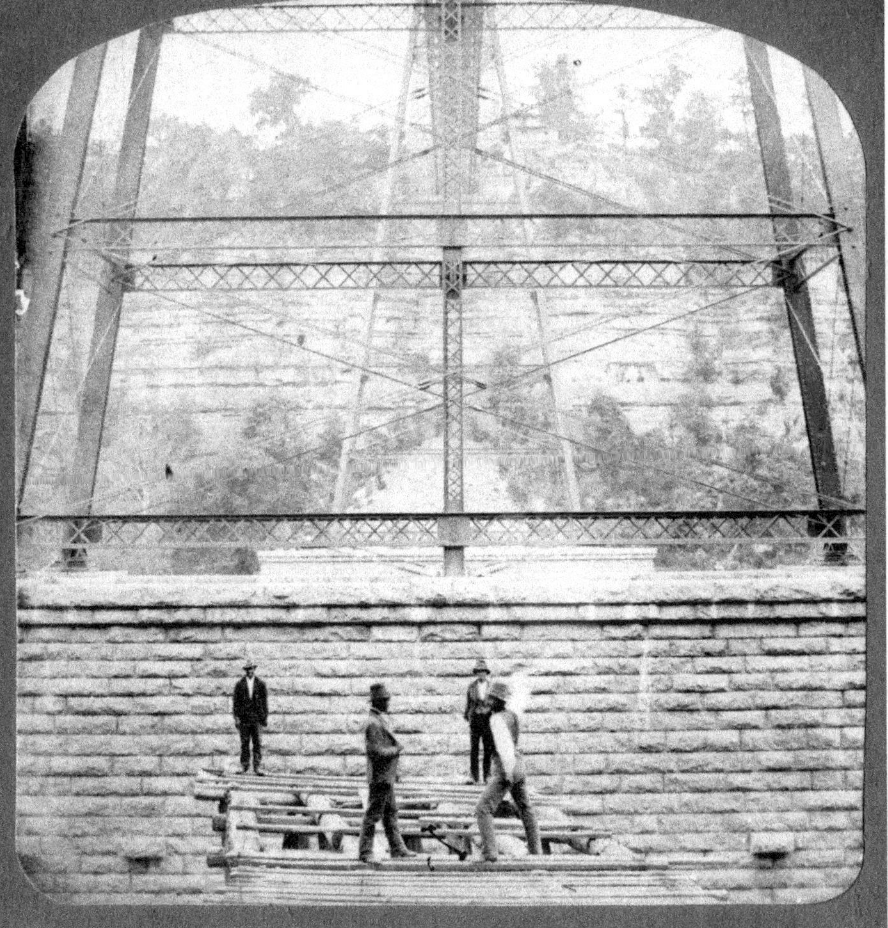

These unidentified bridge workers pose beneath the bridge in this photograph taken in early 1877. They are standing on the scaffolding that was used as a ramp for the large stones of the piers and later to allow easy access for workers to the steel support piers.

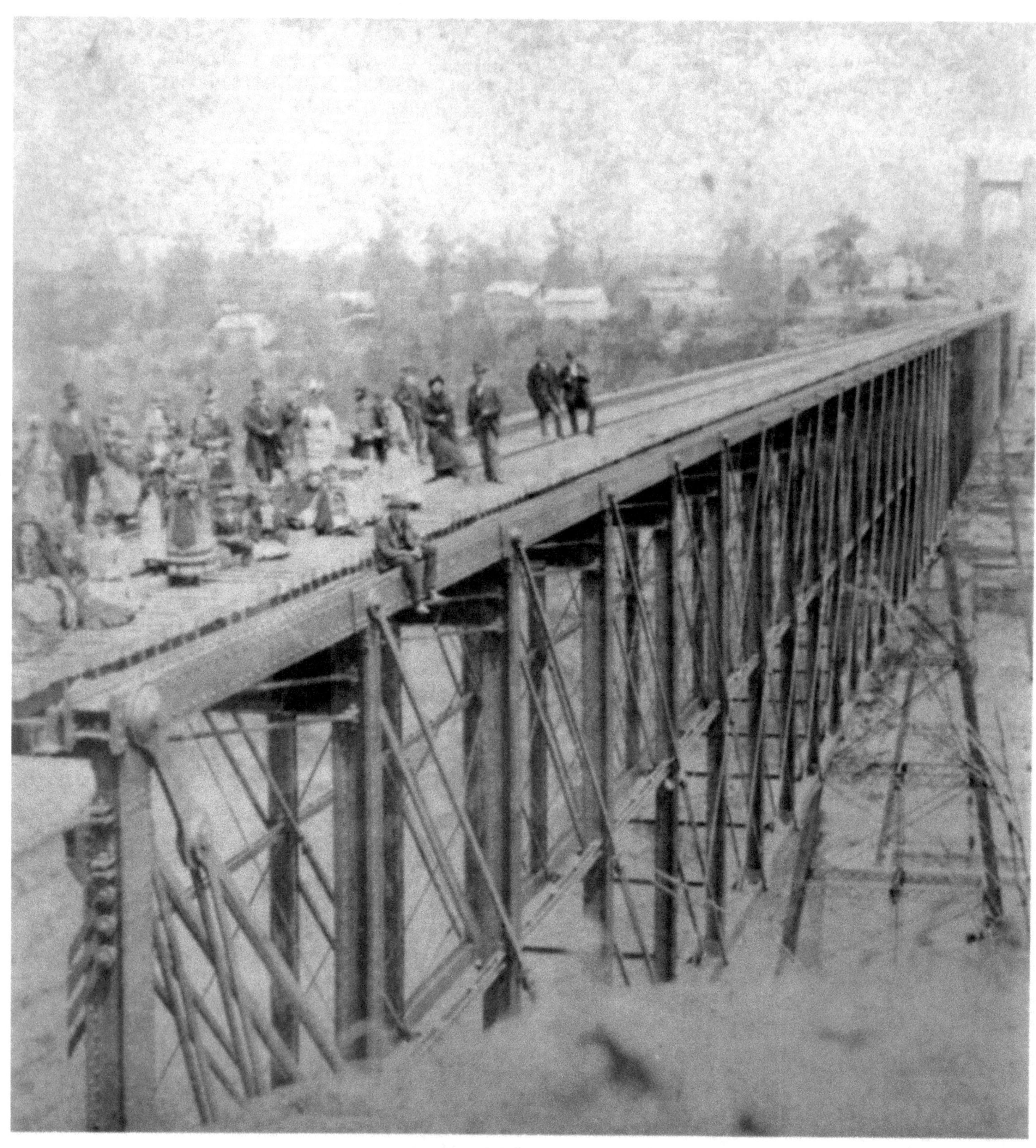

Photographed atop the newly constructed bridge in 1877, these well-dressed sightseers were among the innumerable amounts of people that visited High Bridge in the years to come and posed for an obligatory photograph on the bridge. The photograph was taken from the Jessamine County side looking toward Mercer County.

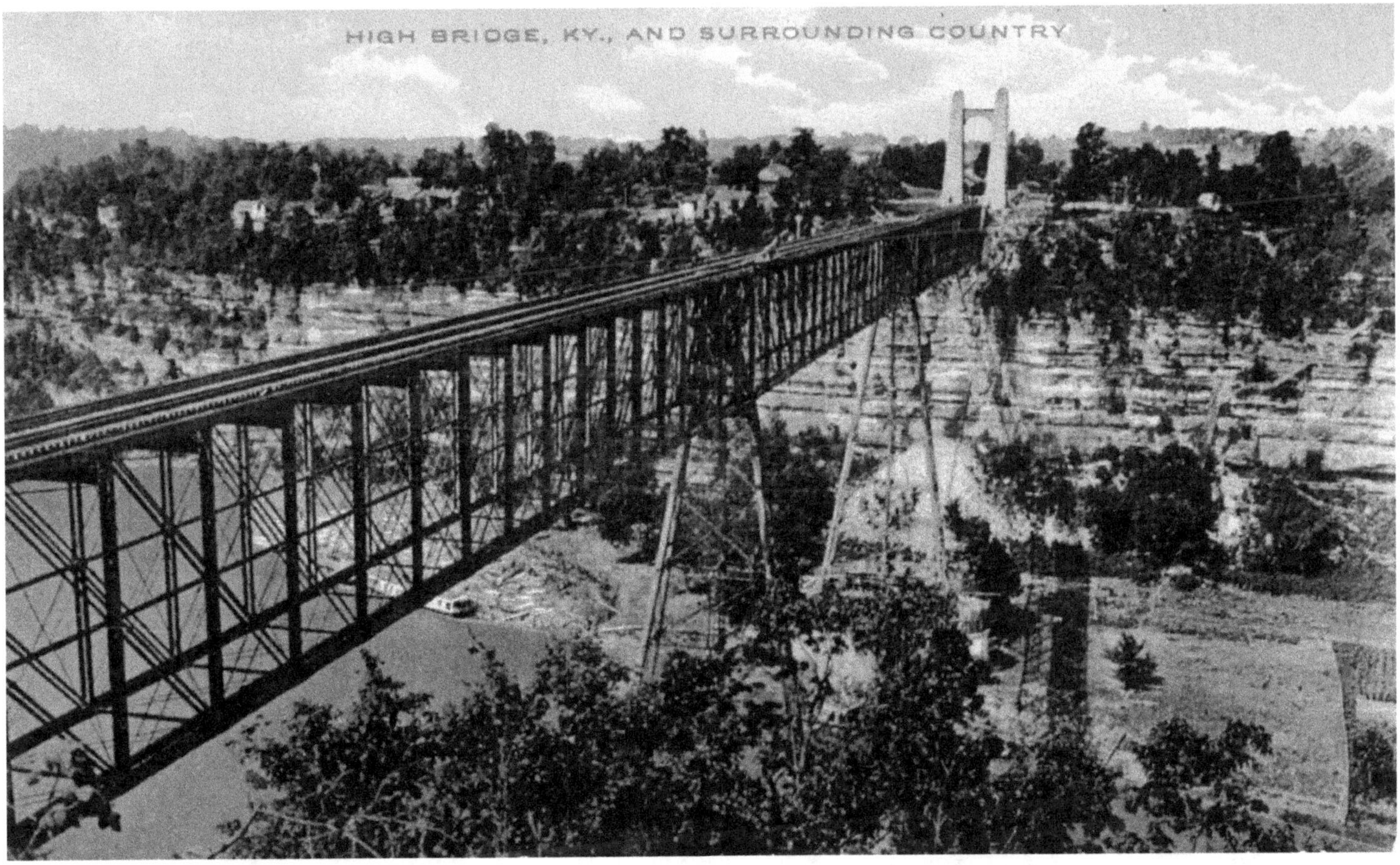

HIGH BRIDGE, KY., AND SURROUNDING COUNTRY

Thousands of people came to High Bridge to observe the testing of the bridge. Many came by trains, which stopped just before the un-tested bridge, allowing the onlookers to disembark and watch from the surrounding cliffs. Once the testing was completed, there was a large celebratory dinner where spirits flowed freely. The *Kentucky Gazette* noted that its reporter dispatched to write about the event was "barely coherent" when he returned to their offices.

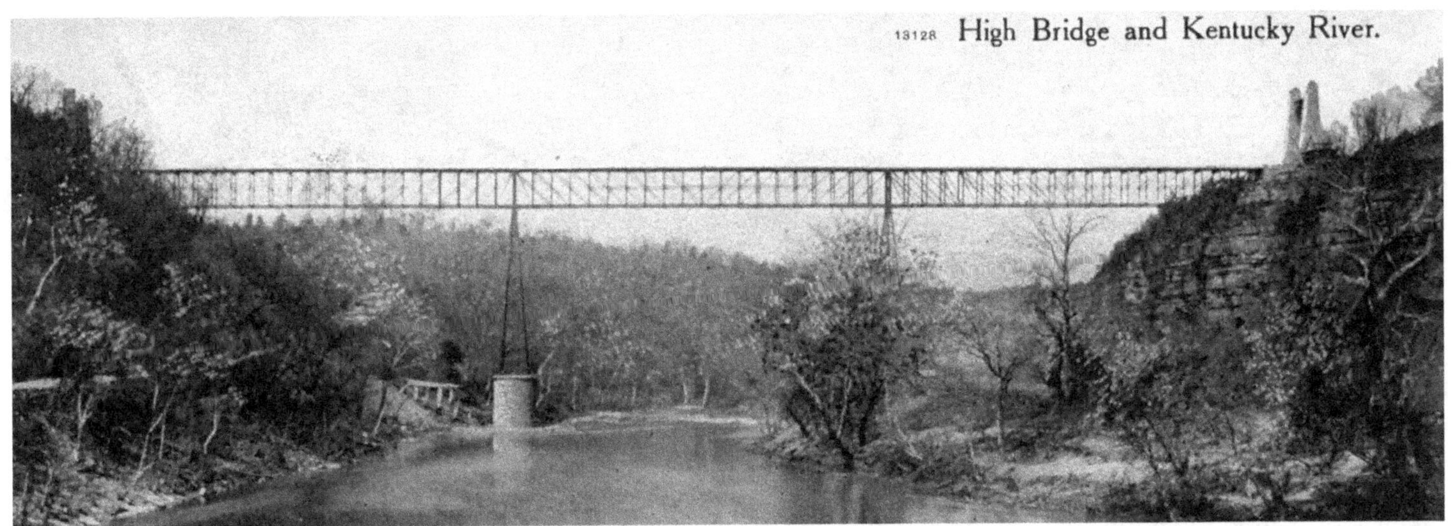

An April 21, 1877, *Kentucky Gazette* article chronicled the testing of the bridge. Four engines were to be used. The first two engines passed slowly over the bridge, then all four engines re-crossed the bridge, meeting in the middle. The final test consisted of twenty-four flatcars, loaded with iron, which were backed across the bridge at twenty miles an hour and brought to a sudden stop. Gov. James McCreary spoke briefly after the tests were concluded.

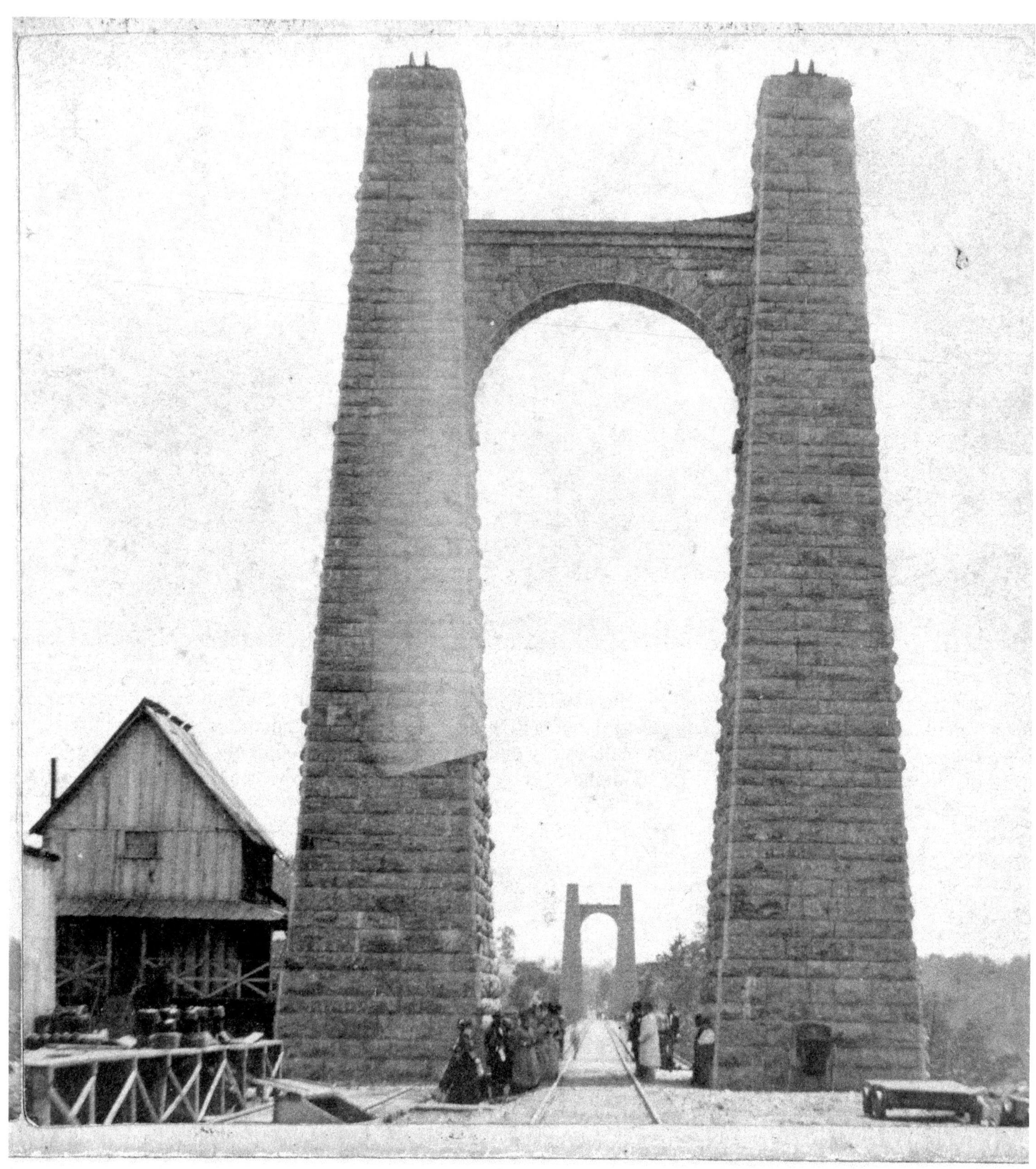

Taken shortly after the bridge was completed, one can see the expanse of the bridge between the two stone Roebling towers. The two-story building to the left of the Jessamine County tower served as an office and for storage during the bridge's construction. It served as the first depot at the site for a short time after the bridge was opened.

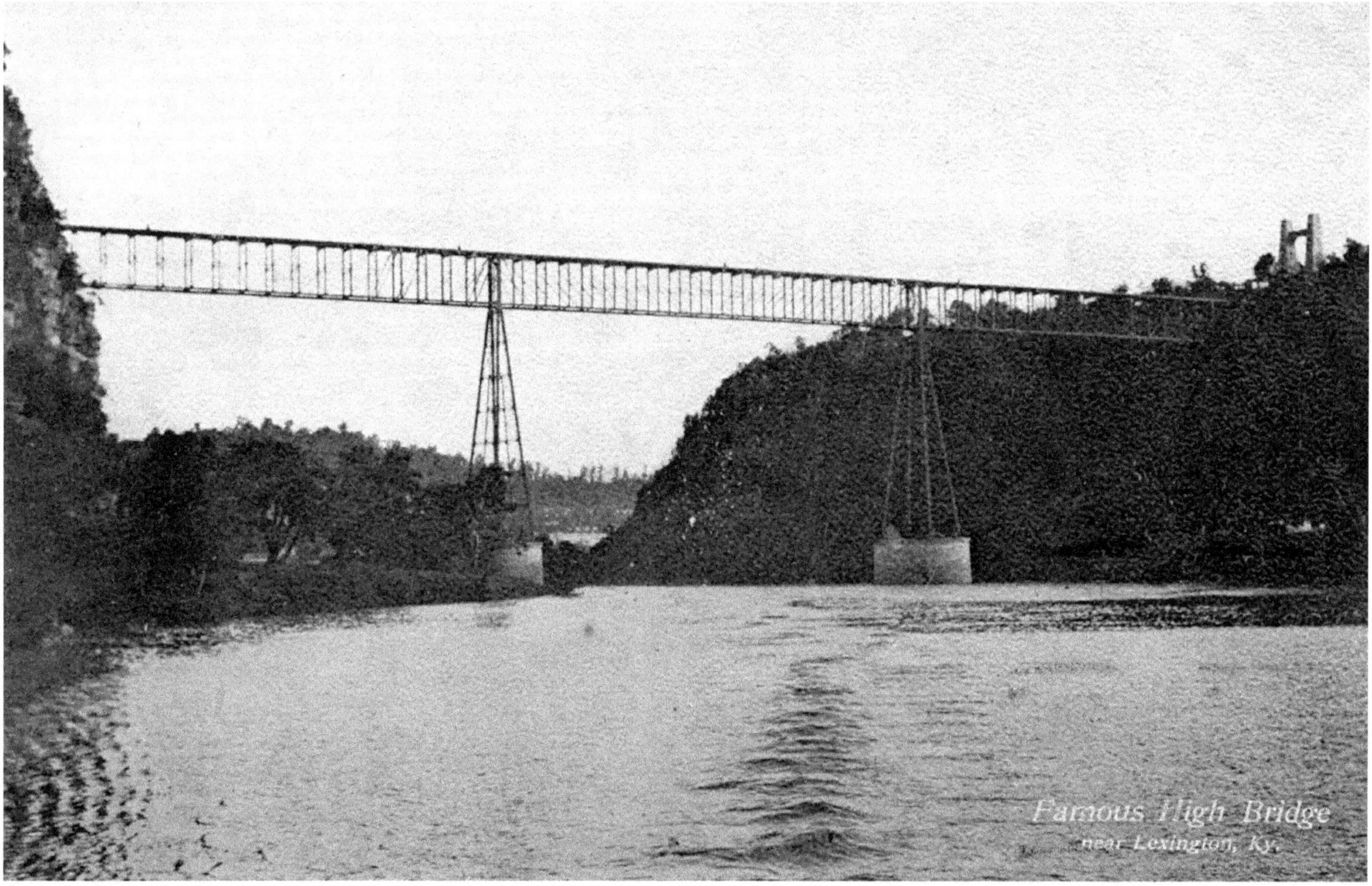

Famous High Bridge
near Lexington, Ky.

Though completed in 1877, the bridge was not formally dedicated until September 17, 1879. The dedication was a lavish event with Pres. Rutherford B. Hayes in attendance as a guest of the Southern Railway System. The public was invited, and they could take the train round-trip from Lexington for 85¢. Despite the fact that it was "raining cats and dogs," according to the *Lexington Press*, the dedication was heavily attended.

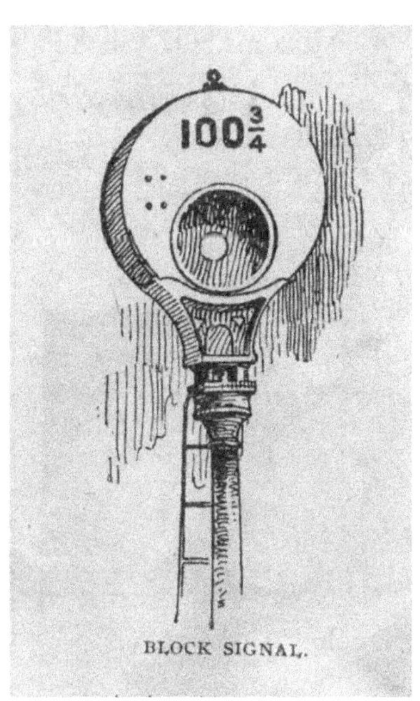

BLOCK SIGNAL.

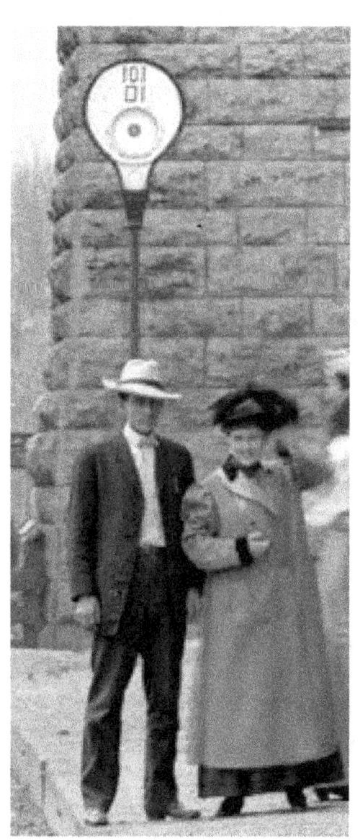

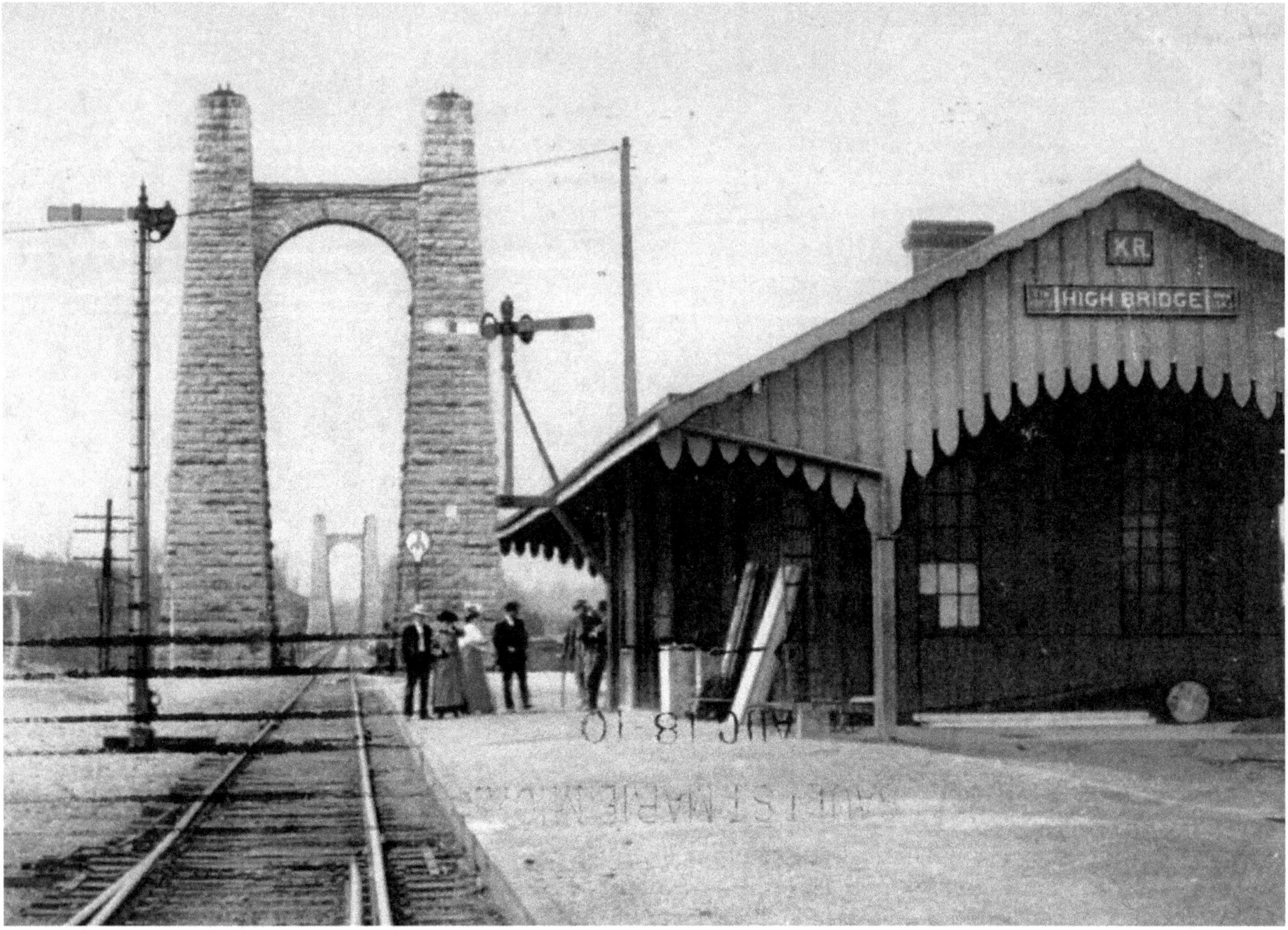

This is the first official depot at High Bridge that greeted passengers. The structure was constructed around 1877 to replace the original two-story building that was used as offices and tool storage during the bridge's construction. This was a common style of depot for the Queen and Crescent Railroad. For example, identical ones were found in Williamstown and in Blanchett, both in Grant County, Kentucky, some seventy miles to the north.

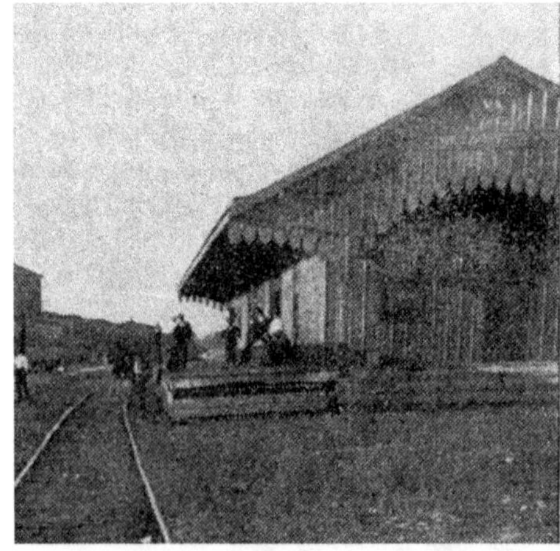

Williamstown Depot.

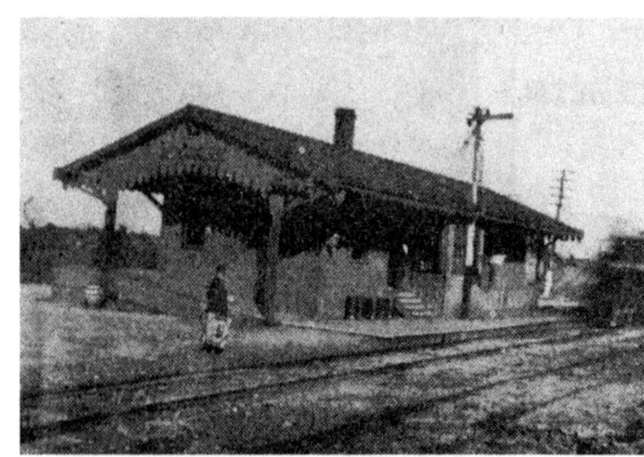

Blanchett Depot.

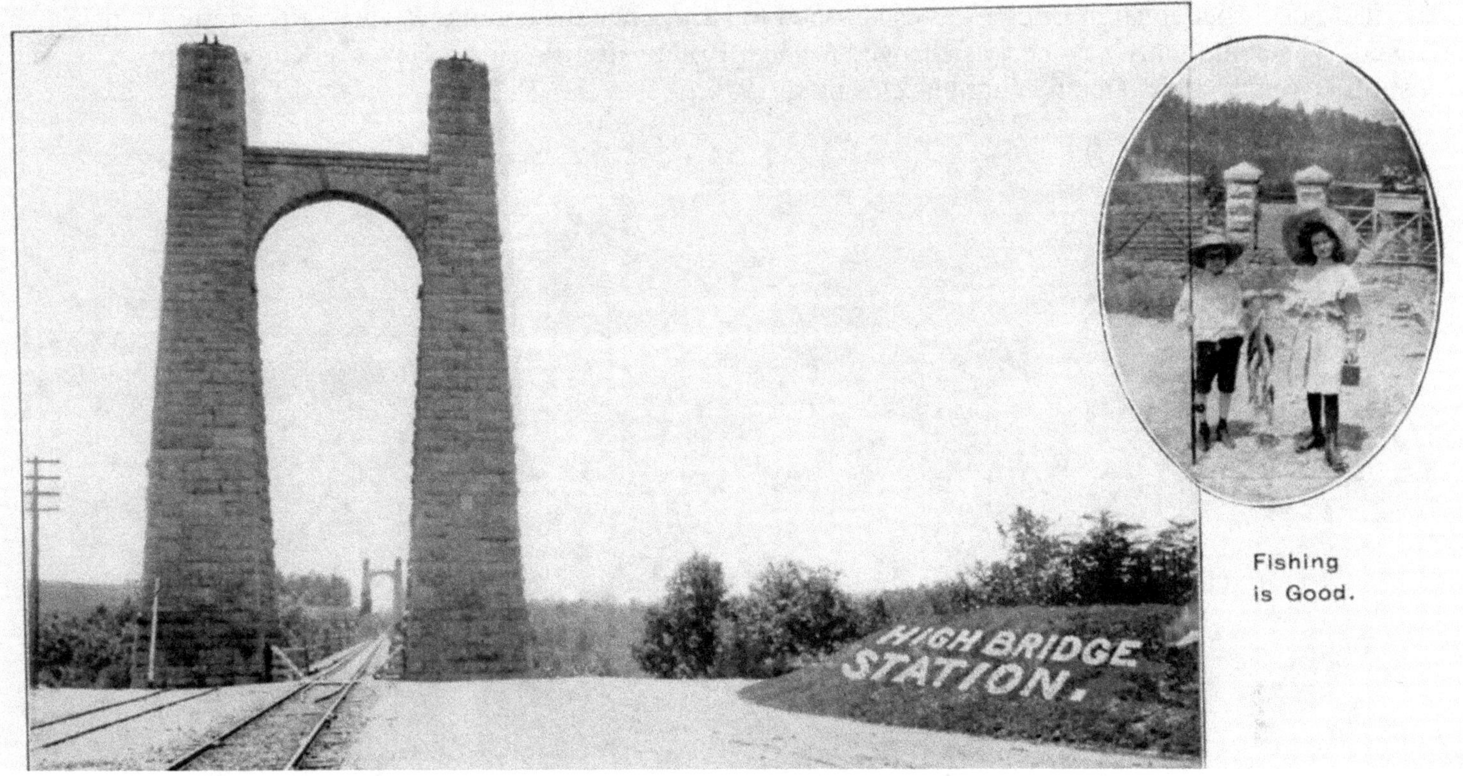

Taken just beyond the ornate Victorian depot, the sense of pride in their bridge and community could be felt by the fact that High Bridge Station is spelled out in flowers in the landscaping just before the landmark towers. The two children displaying their fish in the inset photograph are unidentified, but the photograph was taken outside the entrance to Lock No. 7

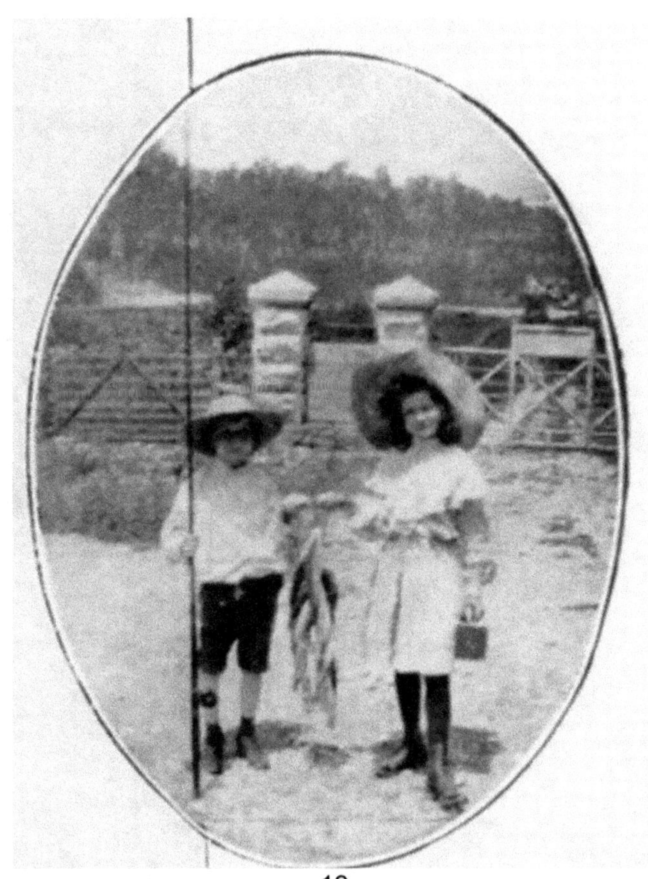

The first post office in High Bridge was established in 1877 when the community was still referred to as North Towers. The community's name was changed to High Bridge after President Hayes's dedication in 1879. The post office remained in operation until its closing in 1976.

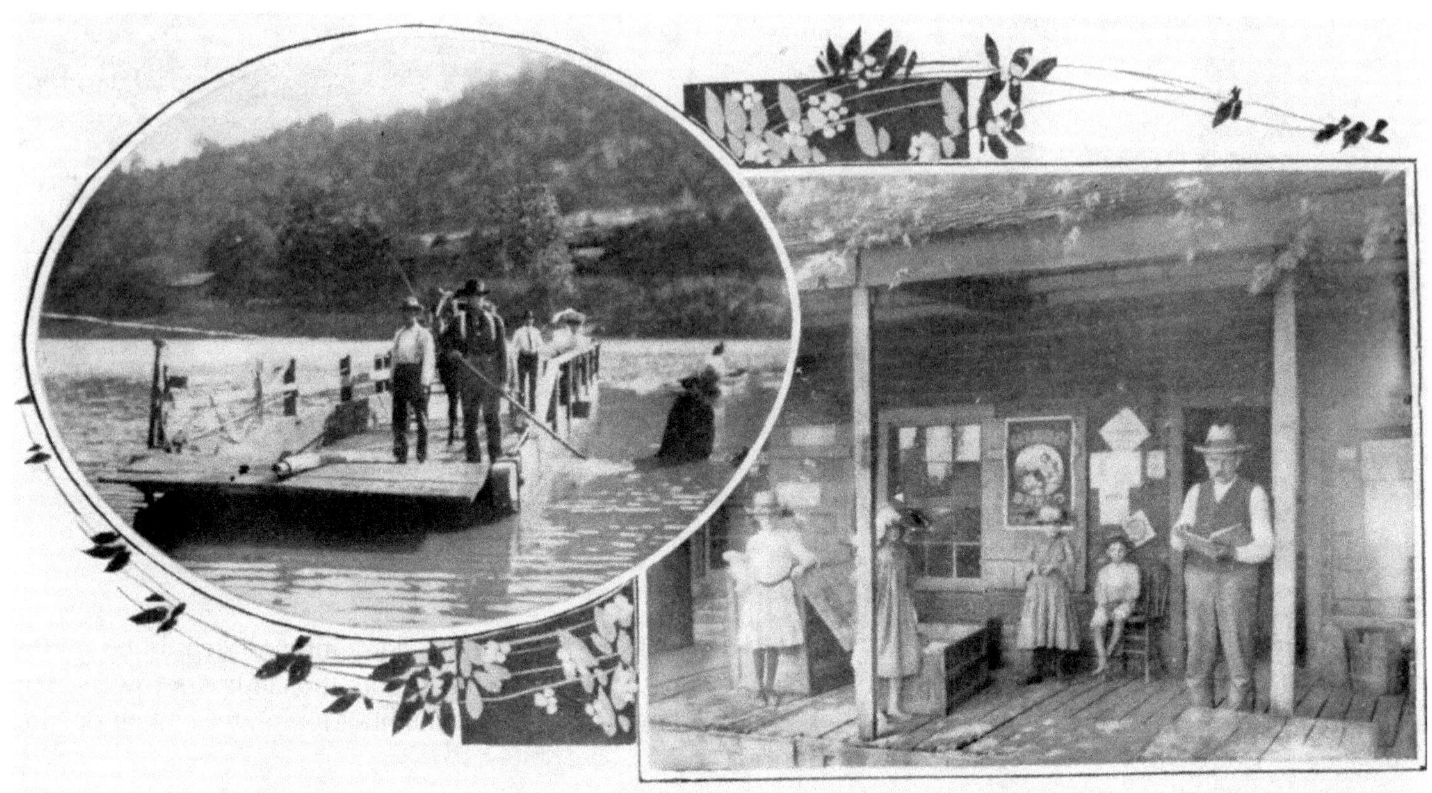

Primitive Ferry at High Bridge.

Country Store and Post Office, High Bridge.

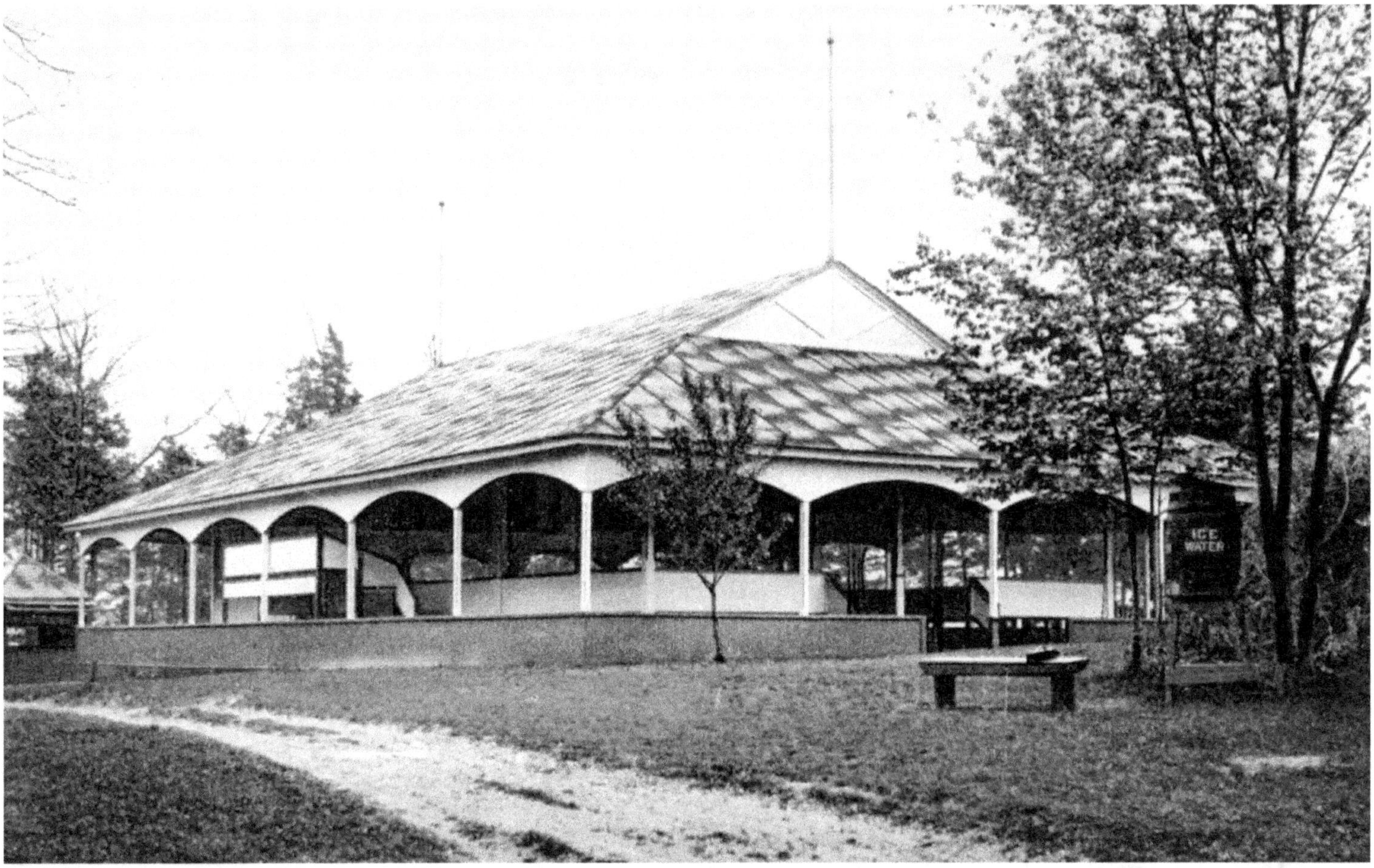

High Bridge Park was created by the High Bridge Association shortly after the bridge's dedication. The centerpiece of the park was the dancing pavilion. Many dances and revivals were held beneath its roof, and if you were a young lady in those days, it was a big deal if a young gentleman asked you to walk the promenade with him just outside the pavilion.

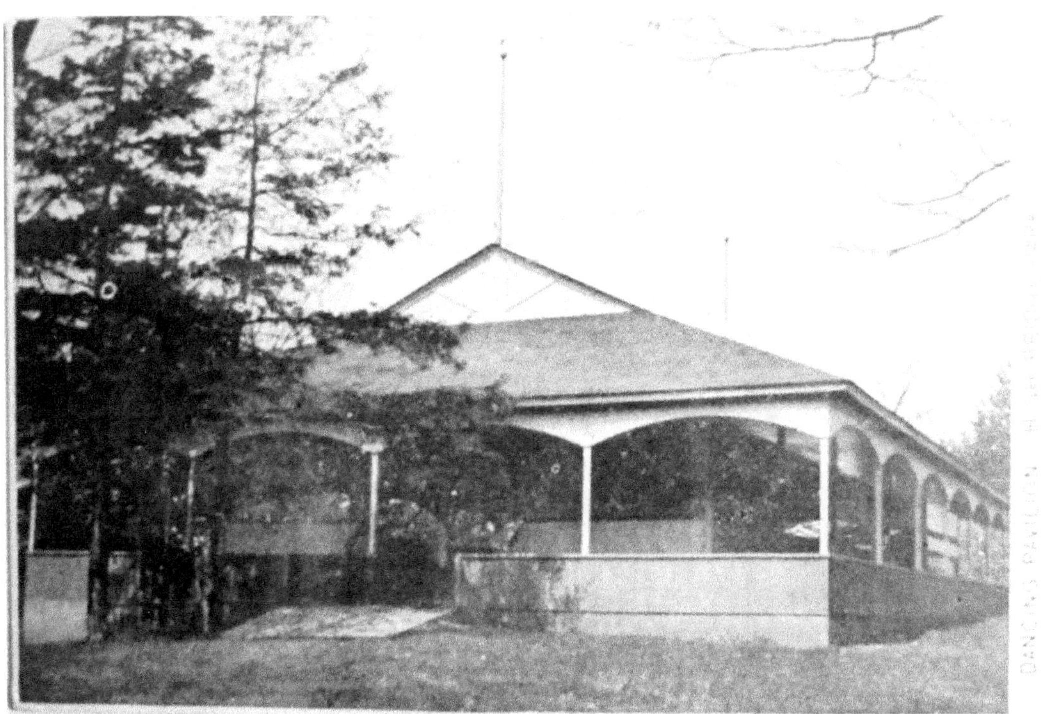

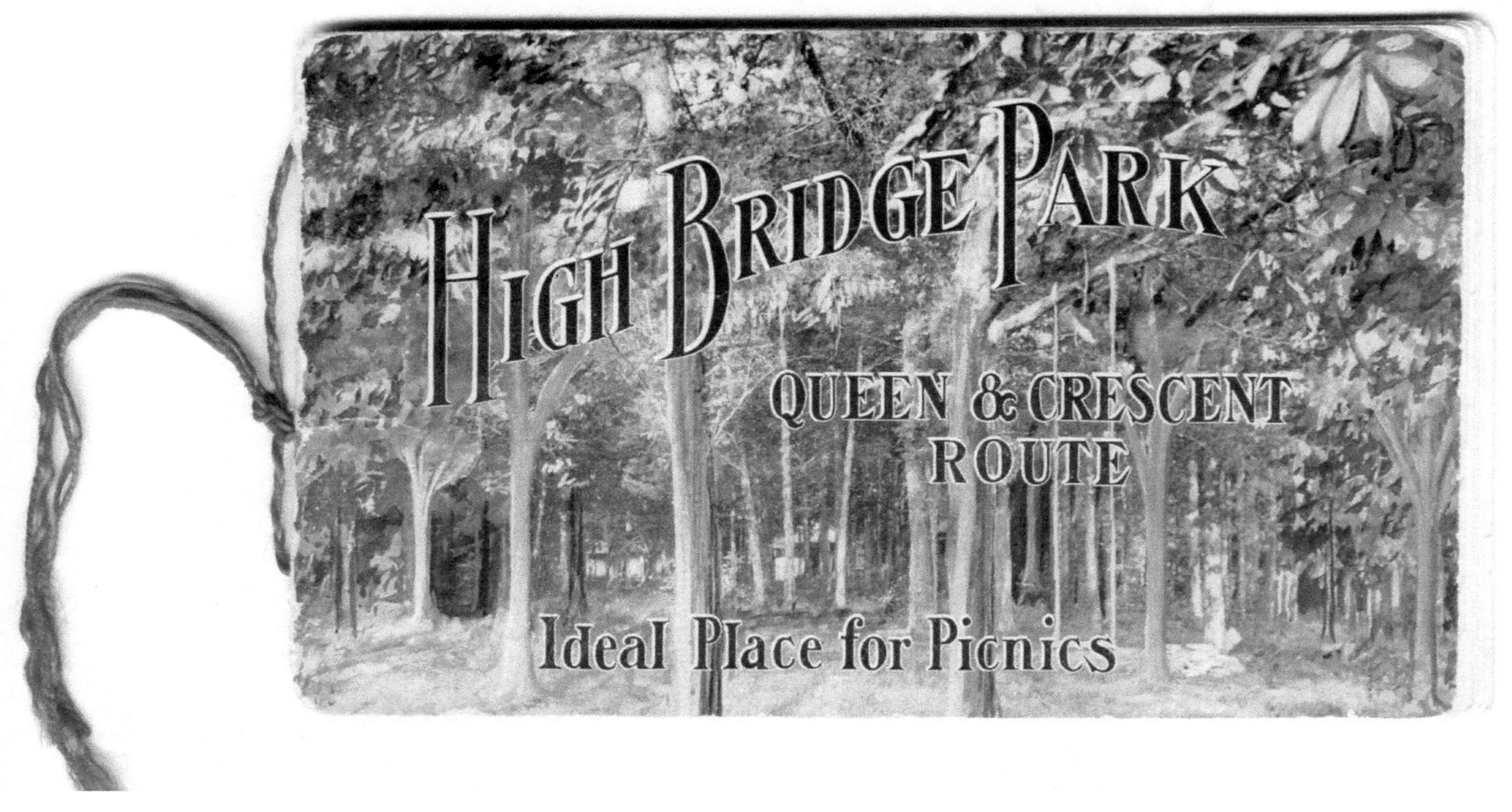

This souvenir booklet, with its exclamation that High Bridge Park was an "ideal place for picnics," was given to excursion passengers by the Queen and Crescent Railroad in the early 1900's. In the binding of the book, a long red string was included so that park visitors could tie the booklet around a button on their shirt or dress and use it as a quick reference guide to the park and not be lost.

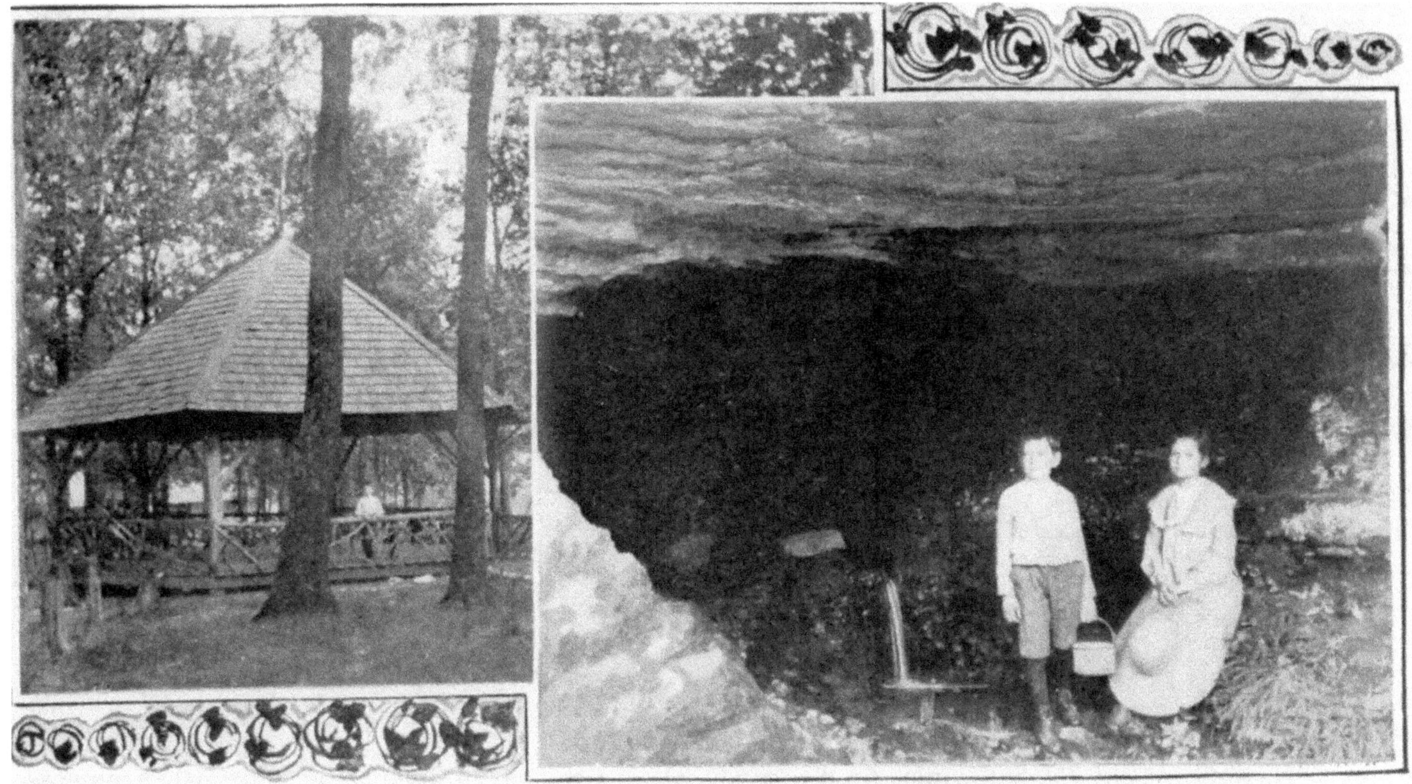

The Pagoda. Cave Spring, High Bridge Park.

Rustic gazebos and walking trails were scattered throughout the park.

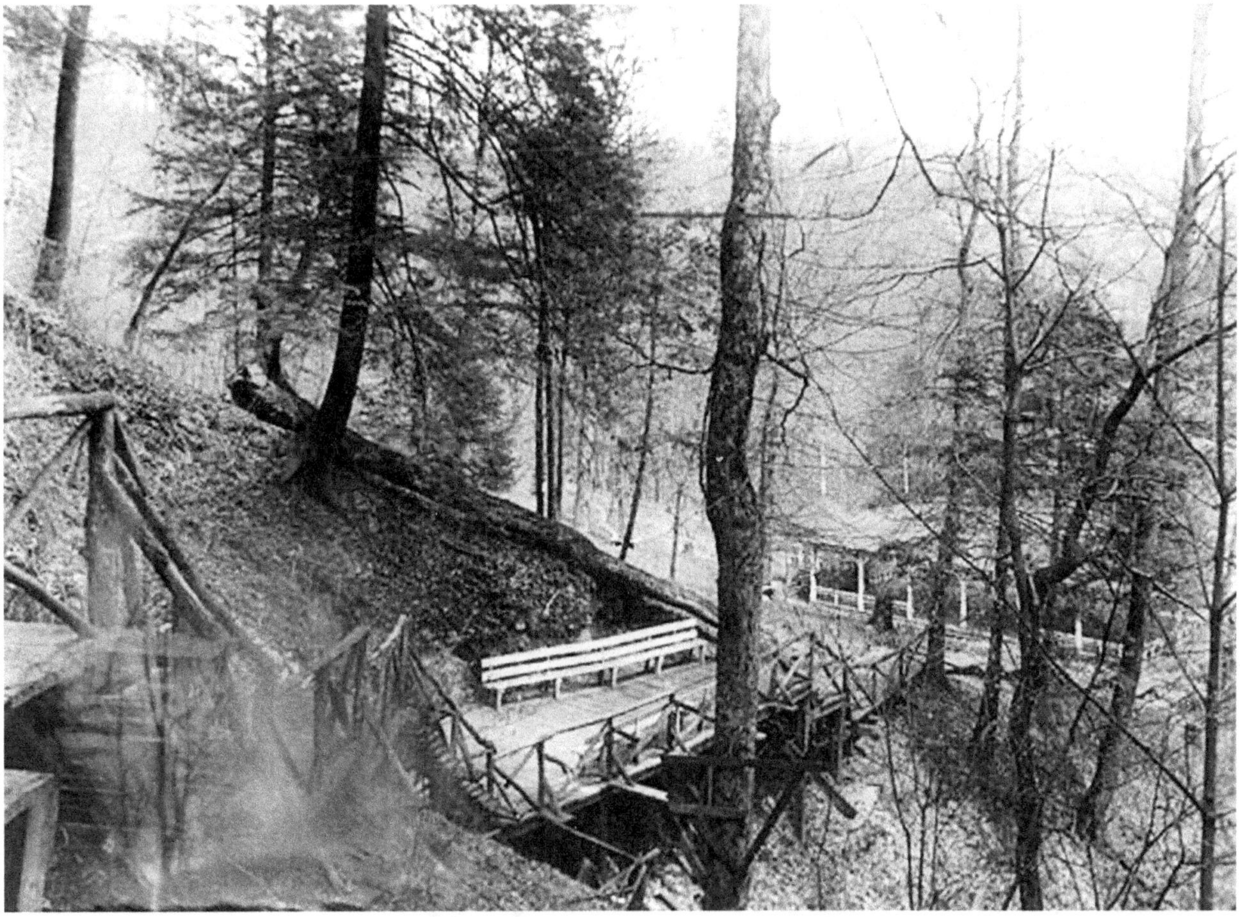

Sir Recd of Mr Joseph Curd My part of the
Surveying fees for 2 Surveys 1 of 7146 acres
and 1 of 5056 Acs by me
Test
Daniel Boone
John Curd
Novem 12th 1792

This receipt is said to be a facsimile of the receipt Daniel Boone gave to John Curd for land surveyed by him in the area on November 12, 1792. In payment for the surveying fee, Curd gave Daniel Boone land that encompassed that used for High Bridge Park in the late Nineteenth Century.

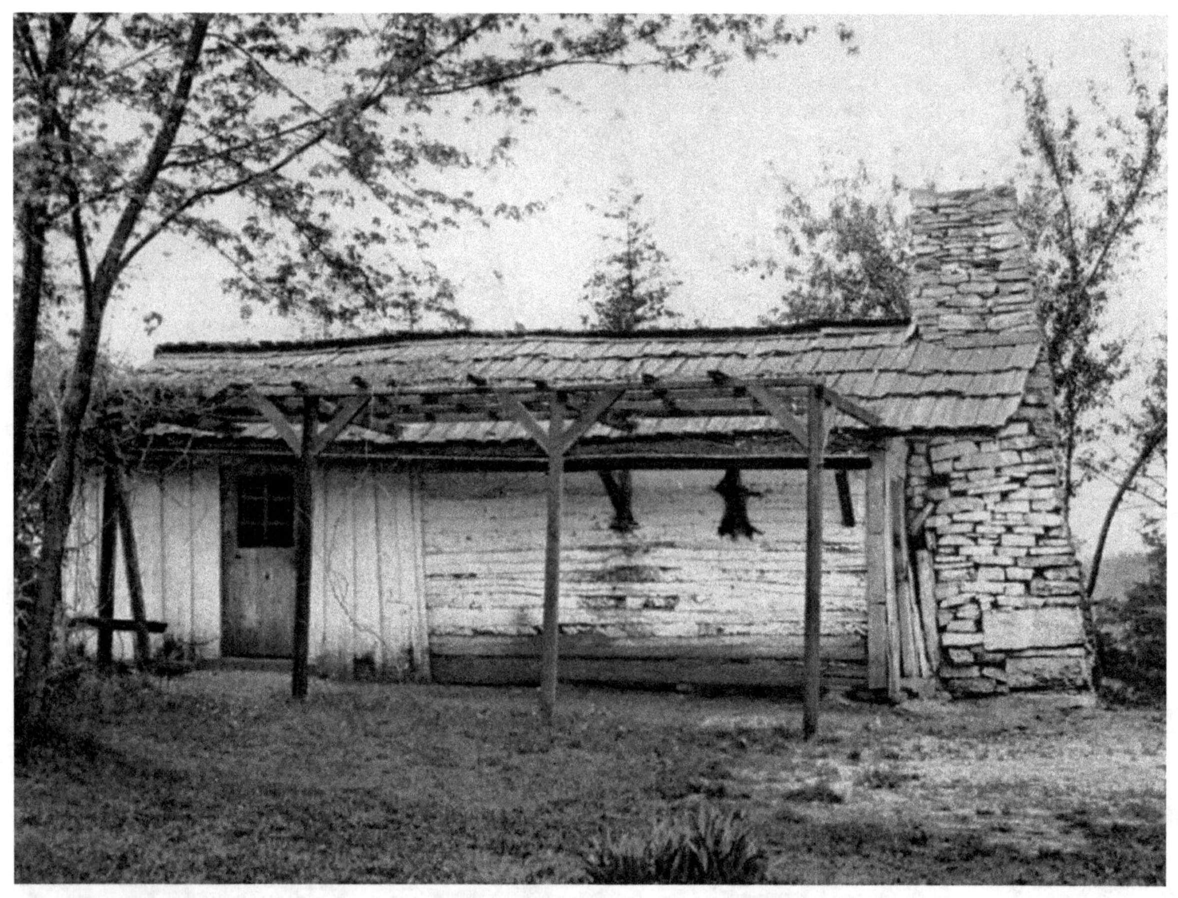

In the years before High Bridge Park, and the bridge itself for that matter, Daniel Boone built a cabin on the cliff edge to stay in on his frequent trips between Harrodsburg and Boonesborough. This photograph was taken of the front of the cabin as it appeared around 1907.

The structure referred to as Boone's Cabin was reconstructed several times over the course of its life, and its appearance altered. In its later years, most of the original cabin had been replaced. Only the stone hearth and a few timbers were from the original Boone-built cabin.

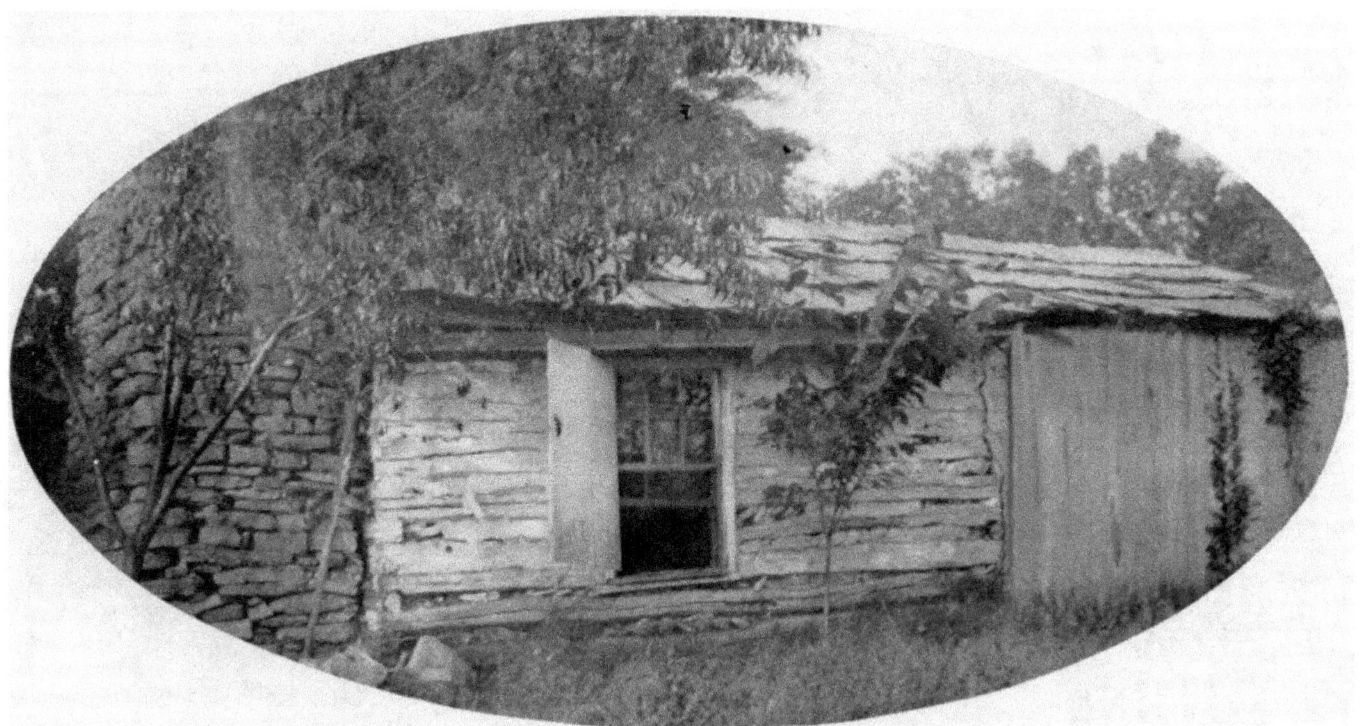

For many years, campers in High Bridge Park could rent Boone's Cabin nightly or weekly for a small fee. The cabin itself was located near the dance pavilion. Today only the foundation of the cabin is visible, if you know where to look.

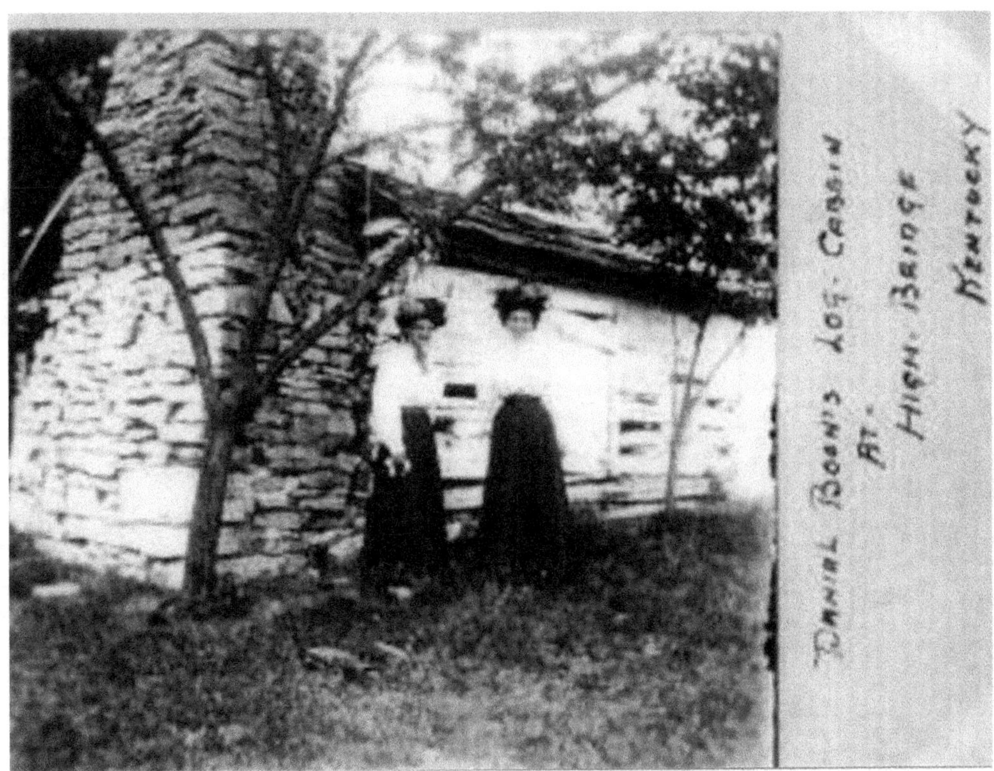

KENTUCKY AND DIX RIVERS, NEAR HIGH BRIDGE, KENTUCKY.

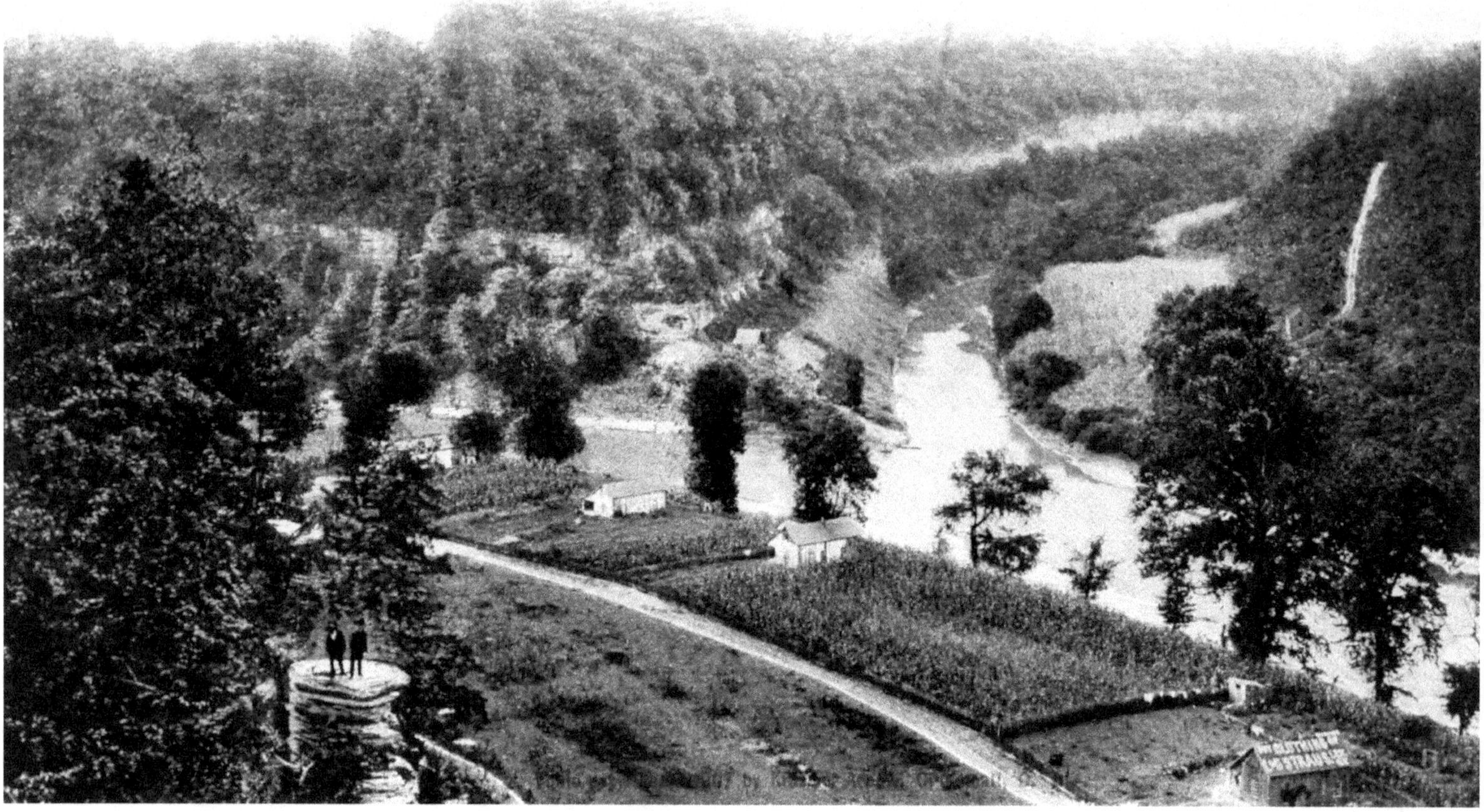

The confluence of the Dix River and the Kentucky River can be seen from High Bridge. The river, originally referred to as Dick's River, was named for a Cherokee Indian chief, Captain Dick, who was friendly with the white men that came to the area around 1769.

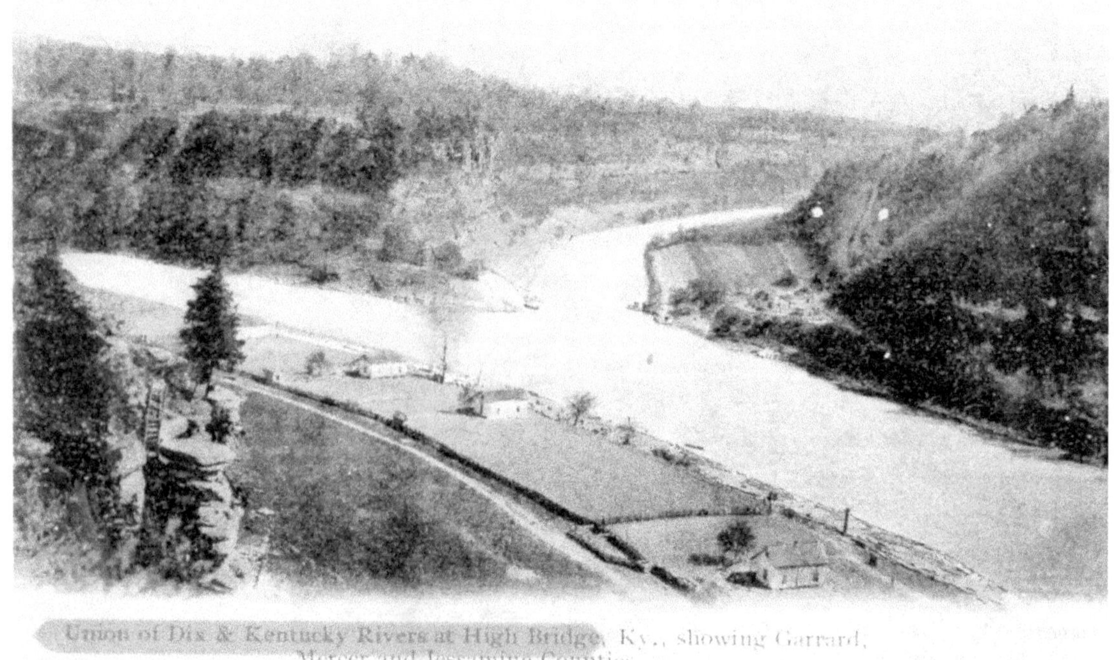

Union of Dix & Kentucky Rivers at High Bridge, Ky., showing Garrard, Mercer and Jessamine Counties.

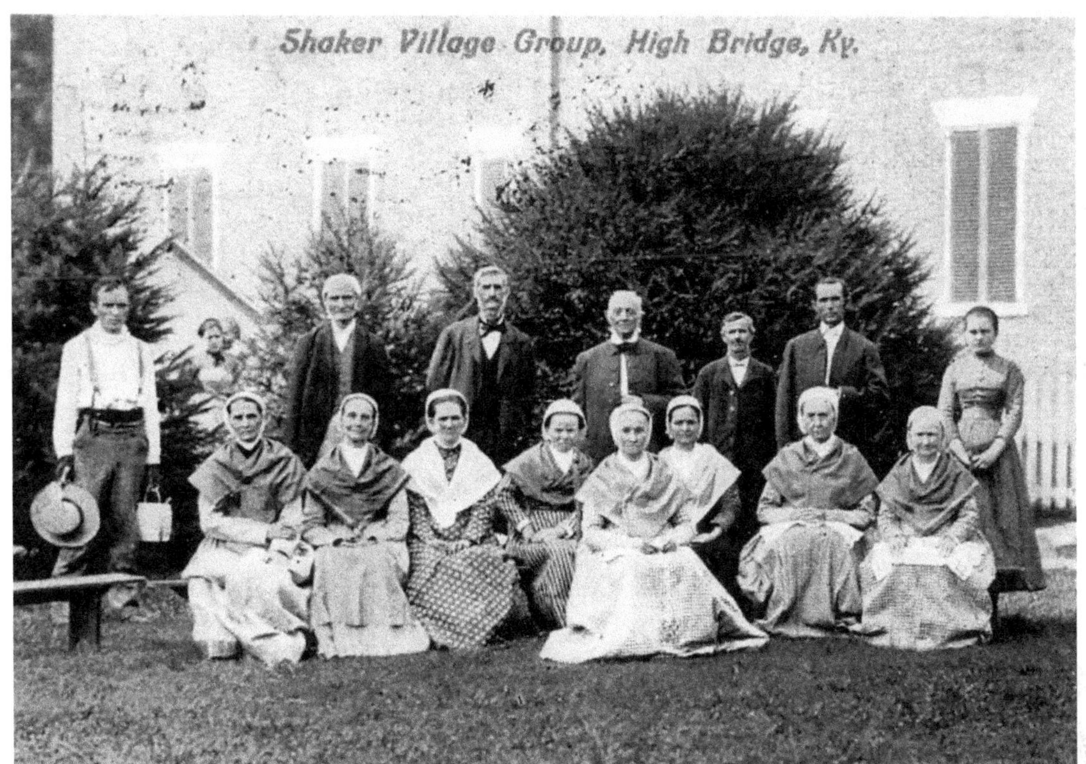

The Society of Believers in Christ's Second Appearing, also known as the Shakers for their religious trance dancing, at one time owned 4,500 acres of land on both sides of the Kentucky River. In their nearby village of Pleasant Hill the Shakers played host to the many land-tourists who came to see High Bridge through the early 1900's.

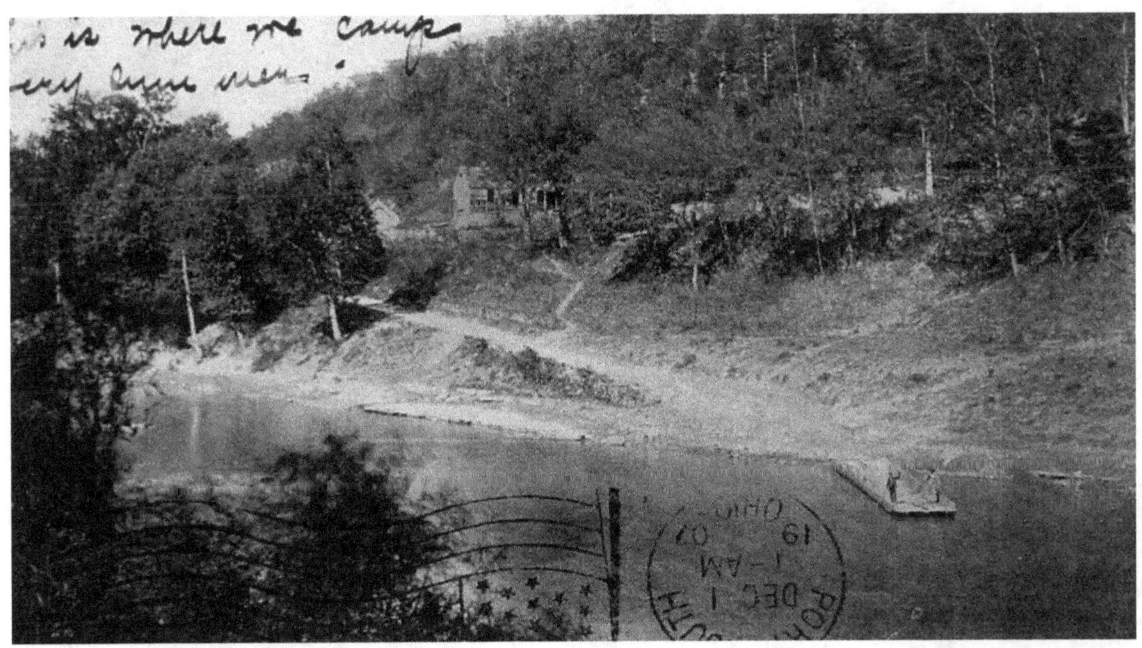

The Shakers bought what is today known as Shaker Ferry Landing from the heirs of Newton Curd. They began operating their ferry in 1830. Starting in 1854, they extensively improved the road down the cliffs in Mercer County to their ferry using dynamite to blast away portions of the cliffs. This work was completed in 1861, which increased the amount of traffic across the river on their ferry.

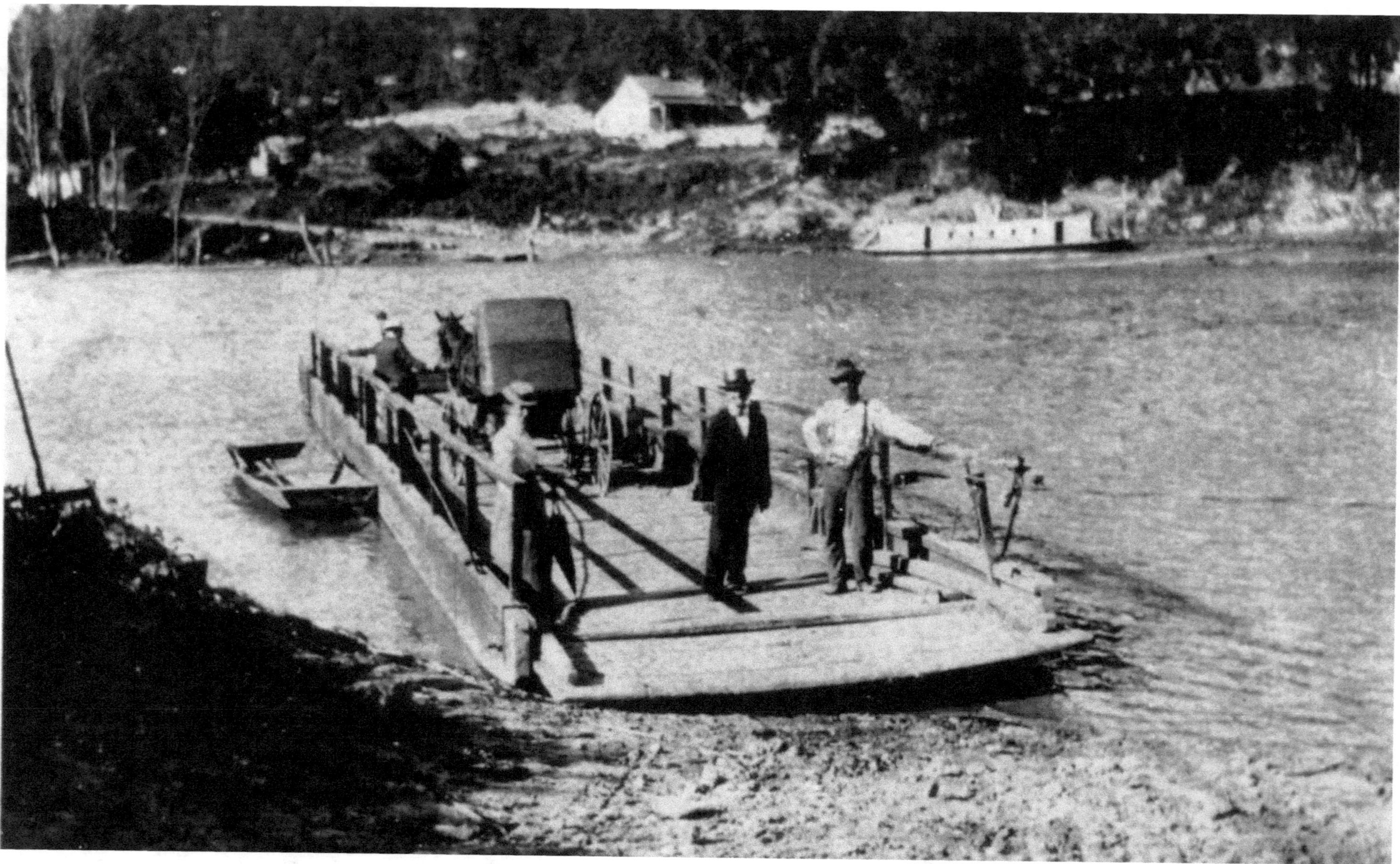

Coming to rest on the Jessamine County side of the river, one can see that the ferry carried foot passengers as well as carriages across the river. Early ferries were propelled across the river by oars, while later ferries were operated by hand-cranked paddlewheels, steam, and eventually diesel-powered engines.

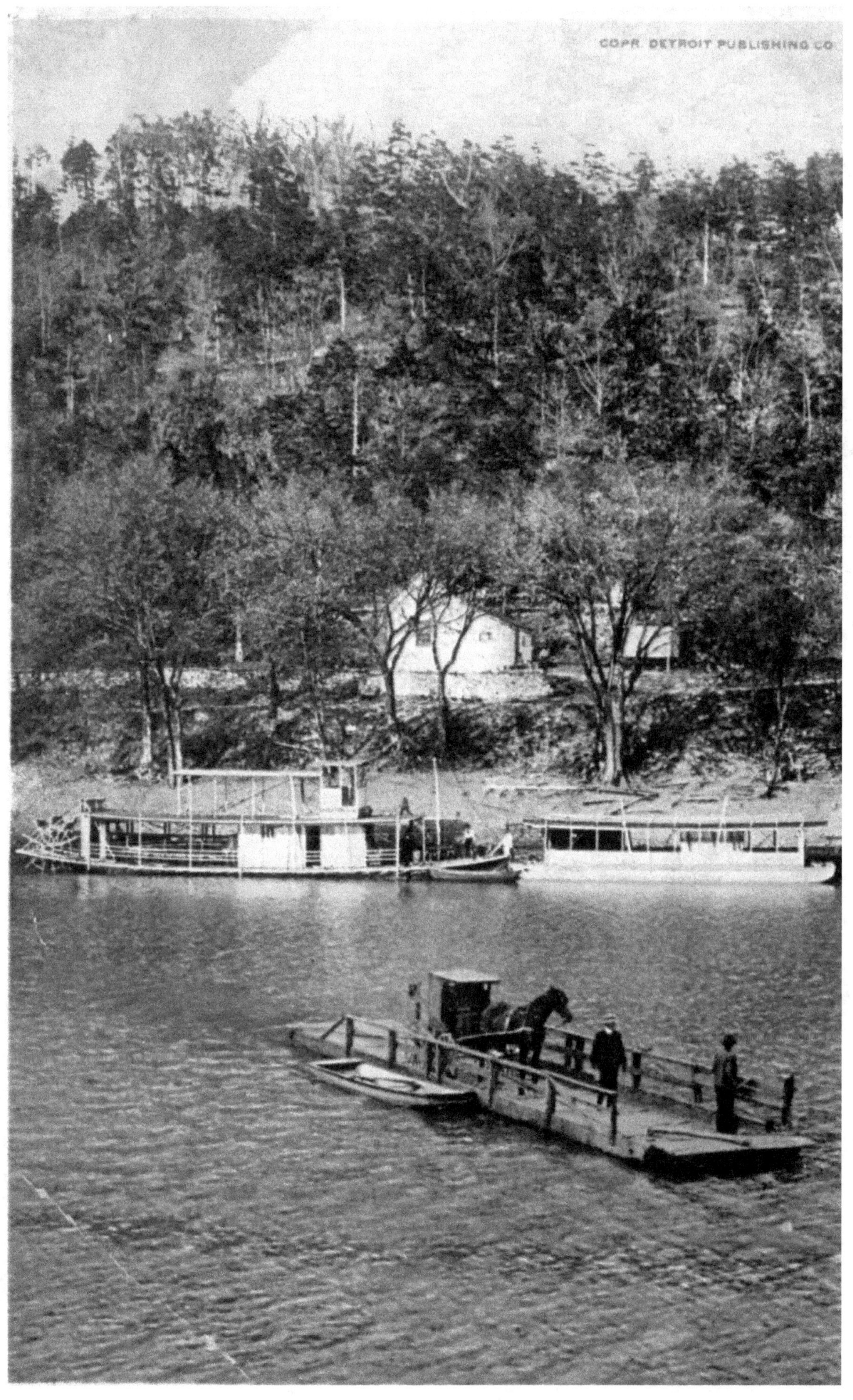
COPR. DETROIT PUBLISHING CO.

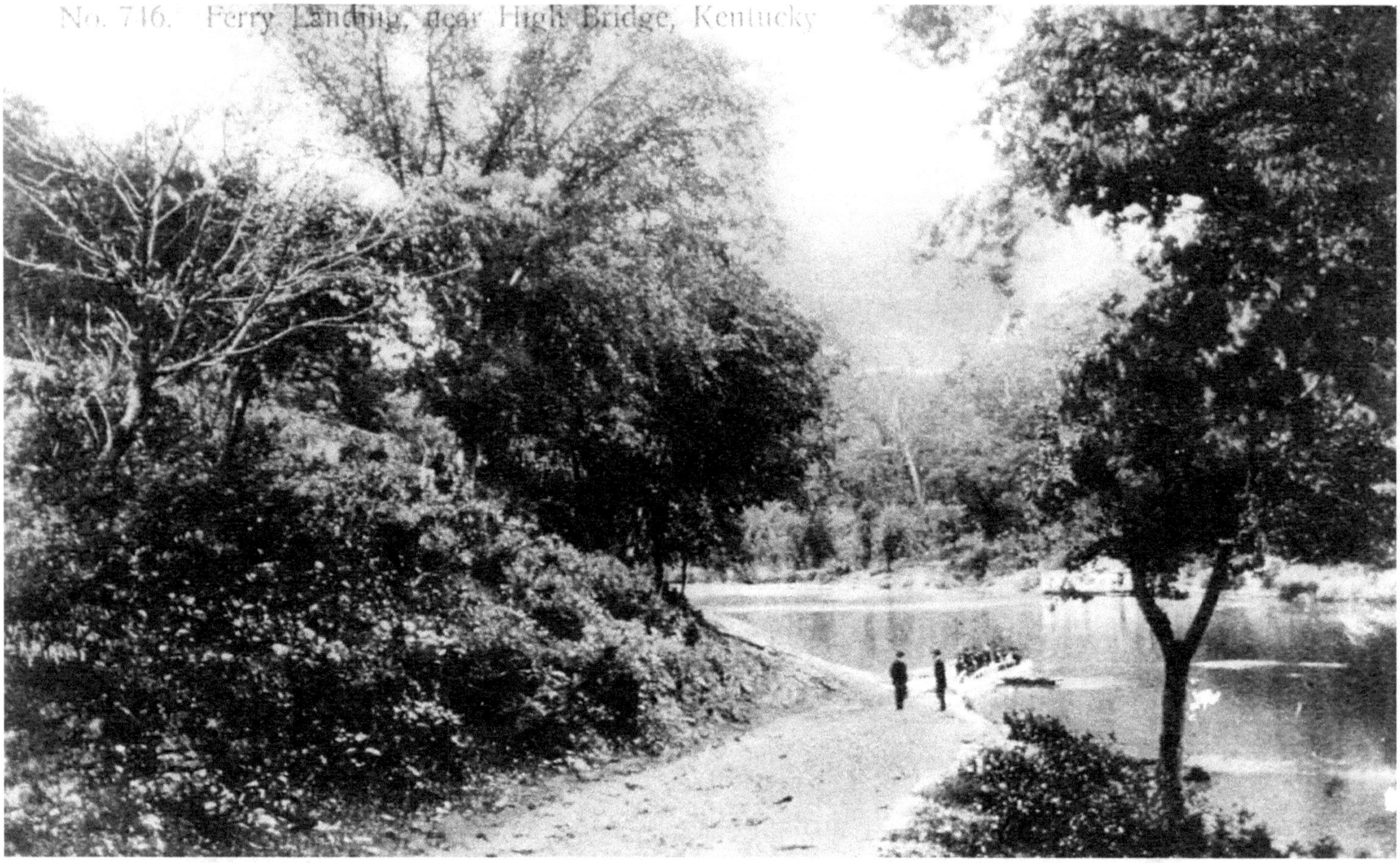

No. 746. Ferry Landing, near High Bridge, Kentucky

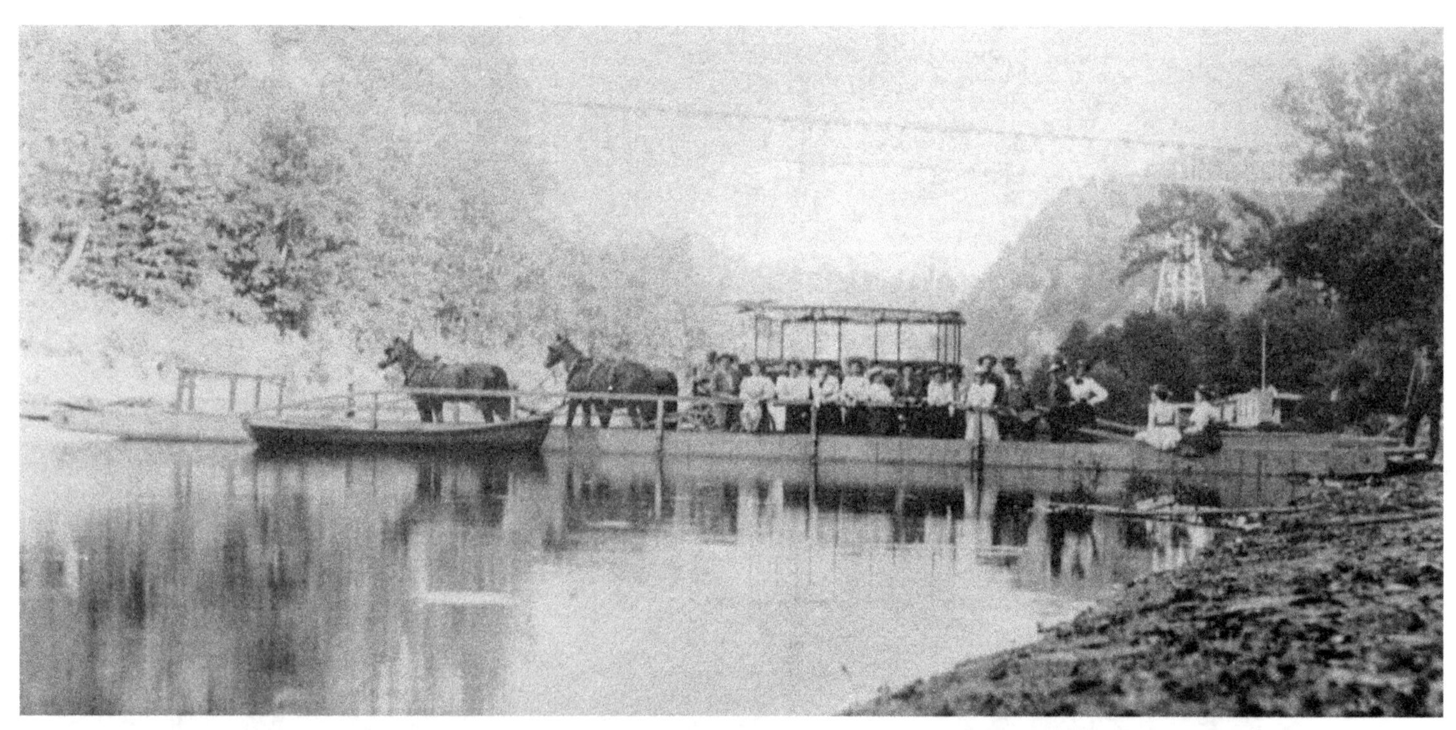

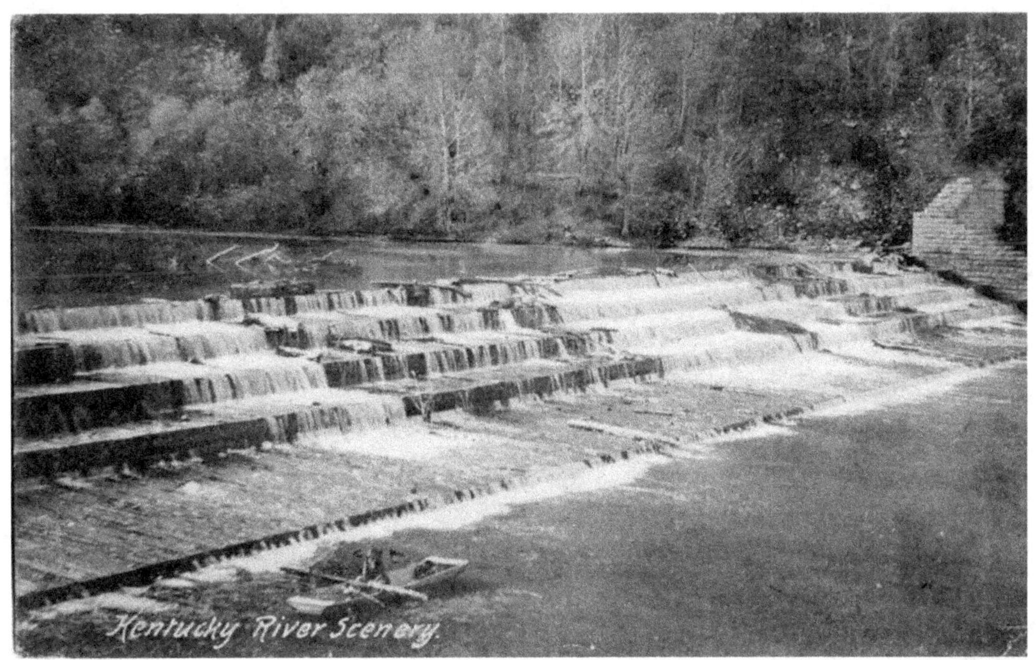

Built in 1896–1897 as part of a project by the Army Corps of Engineers to make the Kentucky River navigable along its entire length, Lock and Dam No. 7 is located near High Bridge and the Shaker landing.

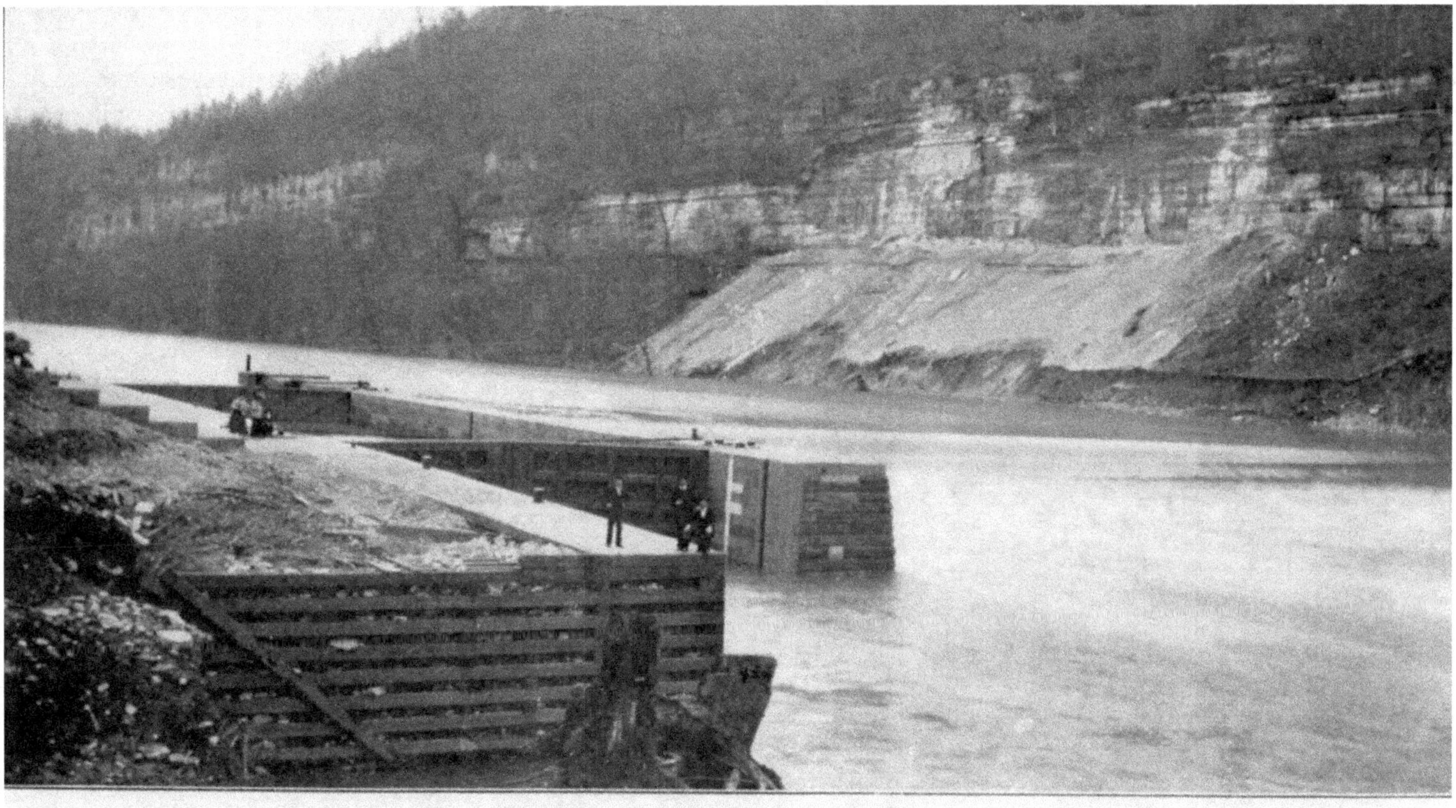

Government Lock, west of High Bridge Park on Kentucky River.

Because of Lock and Dam No. 7, the river was now navigable year-round and not just during a flood tide. In this view of the newly completed lock and dam, on the right hand side of the photograph one can see the pile of silt dredged up from the river and left on its banks in Mercer County. The silt washed away over time.

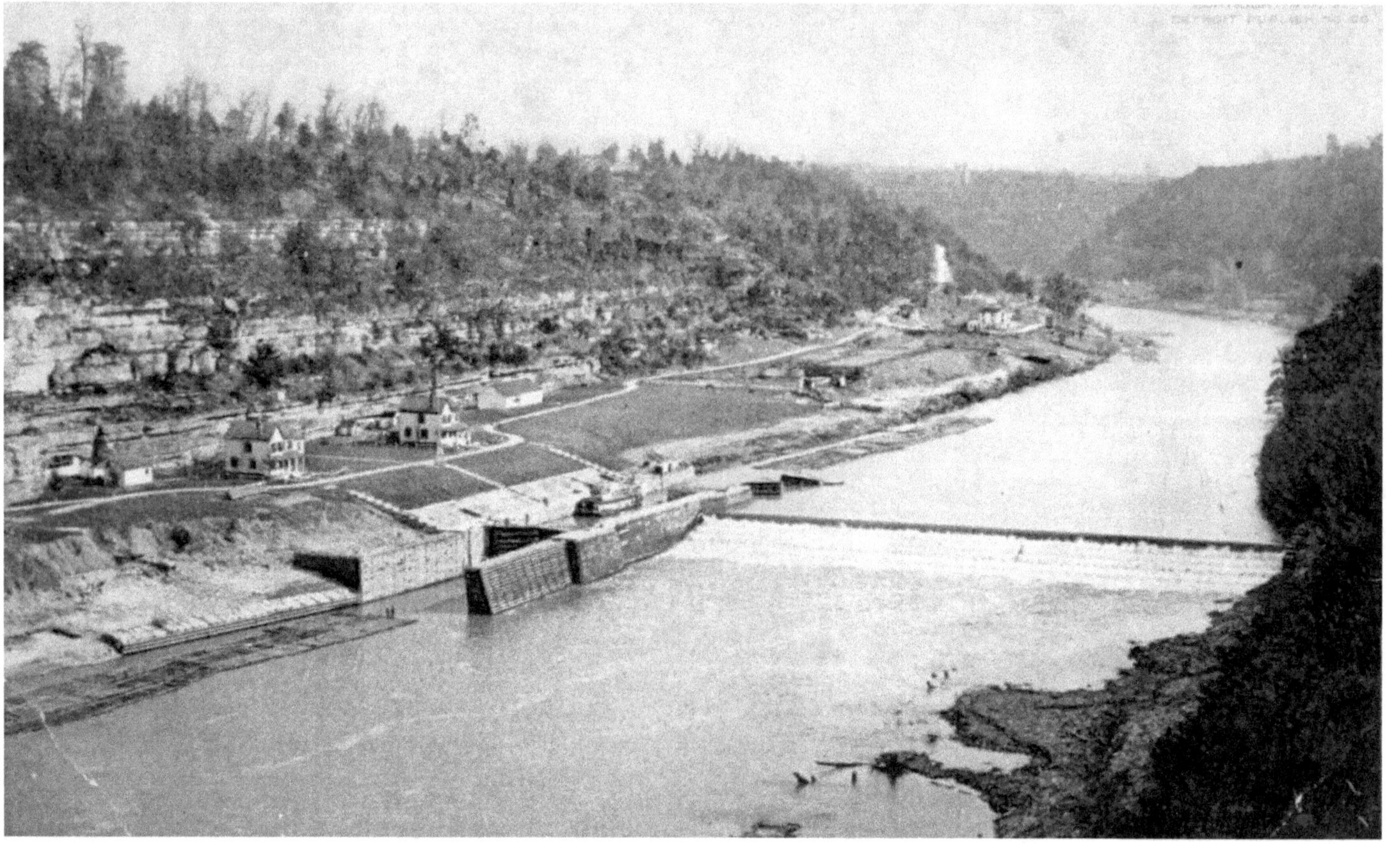

Lock and Dam No. 7 required the Corps of Engineers to purchase land from families and businesses alike due to the fact that the water table would be raised. In this view of the lock, one can see new homes that were built along the river, including the home of the lockmaster.

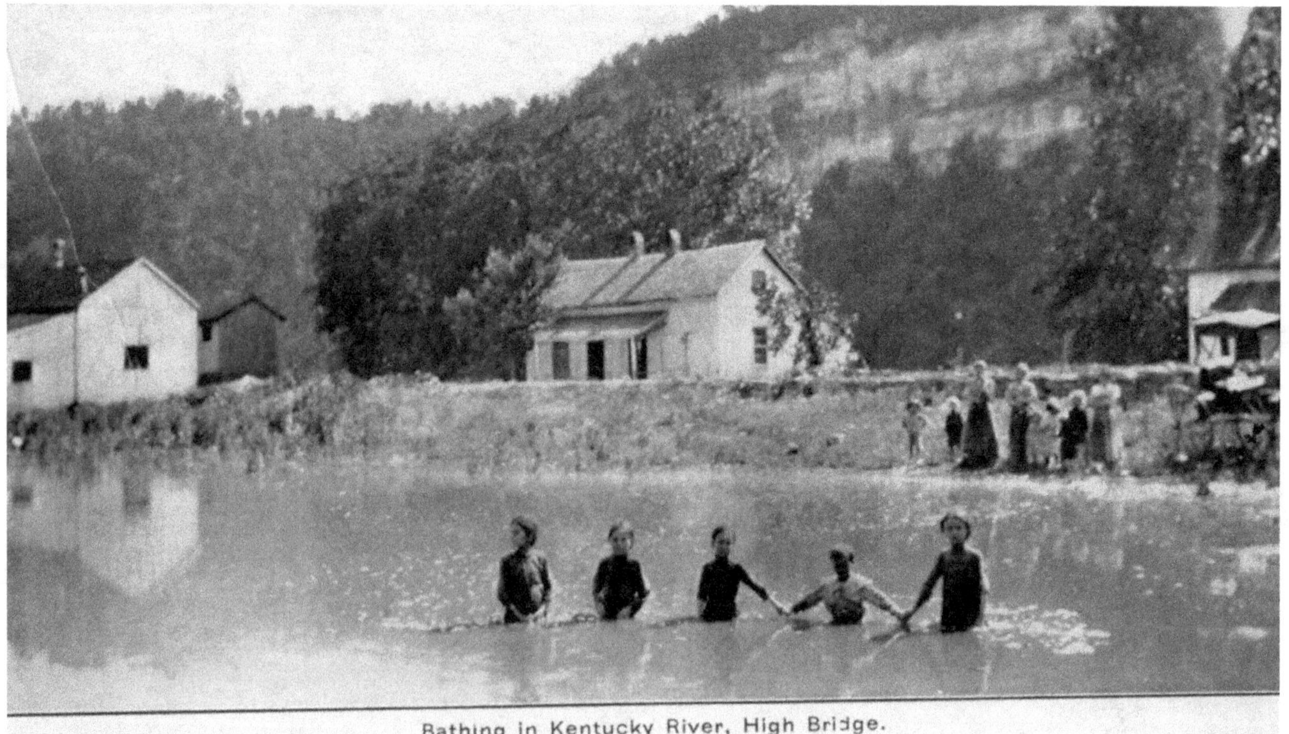

This photograph of the Stephens' farm was taken before Lock and Dam No. 7 were constructed. It demonstrates the shallowness of the riverbed as the children are in the middle of the river and the water is only waist-high. The family relocated before construction began because after the lock and dam were completed most of their farm would be under water

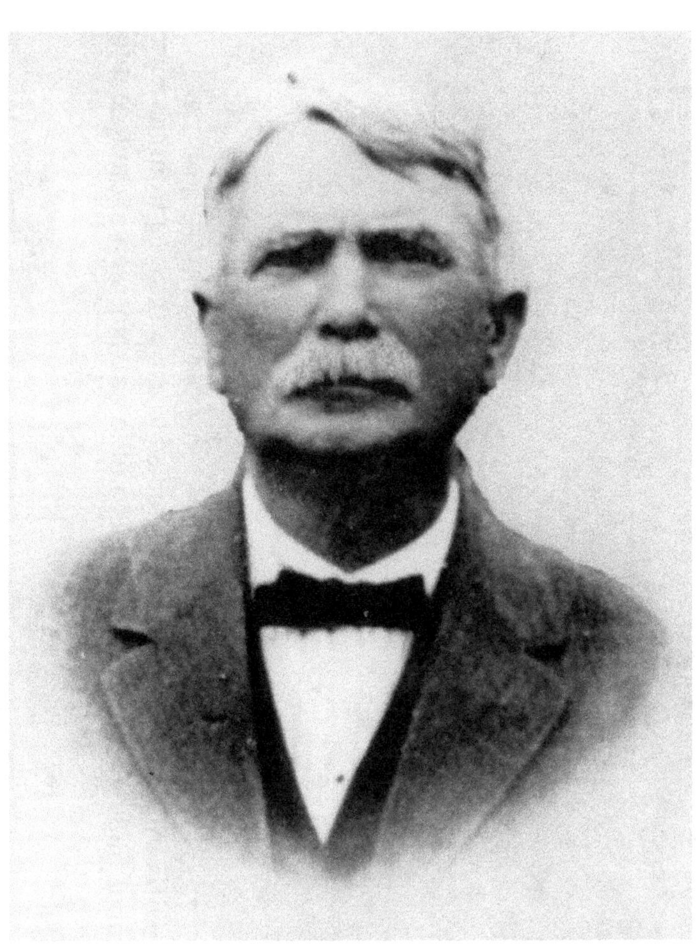

Jeddah D. "Dug" Hughes owned and operated the High Bridge Lumber Company with business partner Willies Alcorn. The sawmill site would be displaced because of the construction of Lock No. 7. However, Hughes did not want to sell out to the Corps of Engineers. Negotiations ensued to no avail, and the land the High Bridge Lumber Company occupied became the only land the corps had to condemn by exercising eminent domain, forcing Hughes to relocate the mill

Logging was once a big business on the Kentucky River. Loggers from as far away as Barbourville would float their logs in rafts to be sold to sawmills. The logs were branded for identification before they made their journey, but occasionally logs were missed. People along the river would search for unbranded logs, claim them, and sell them. The High Bridge Lumber Company was the first sawmill loggers would encounter on their route.

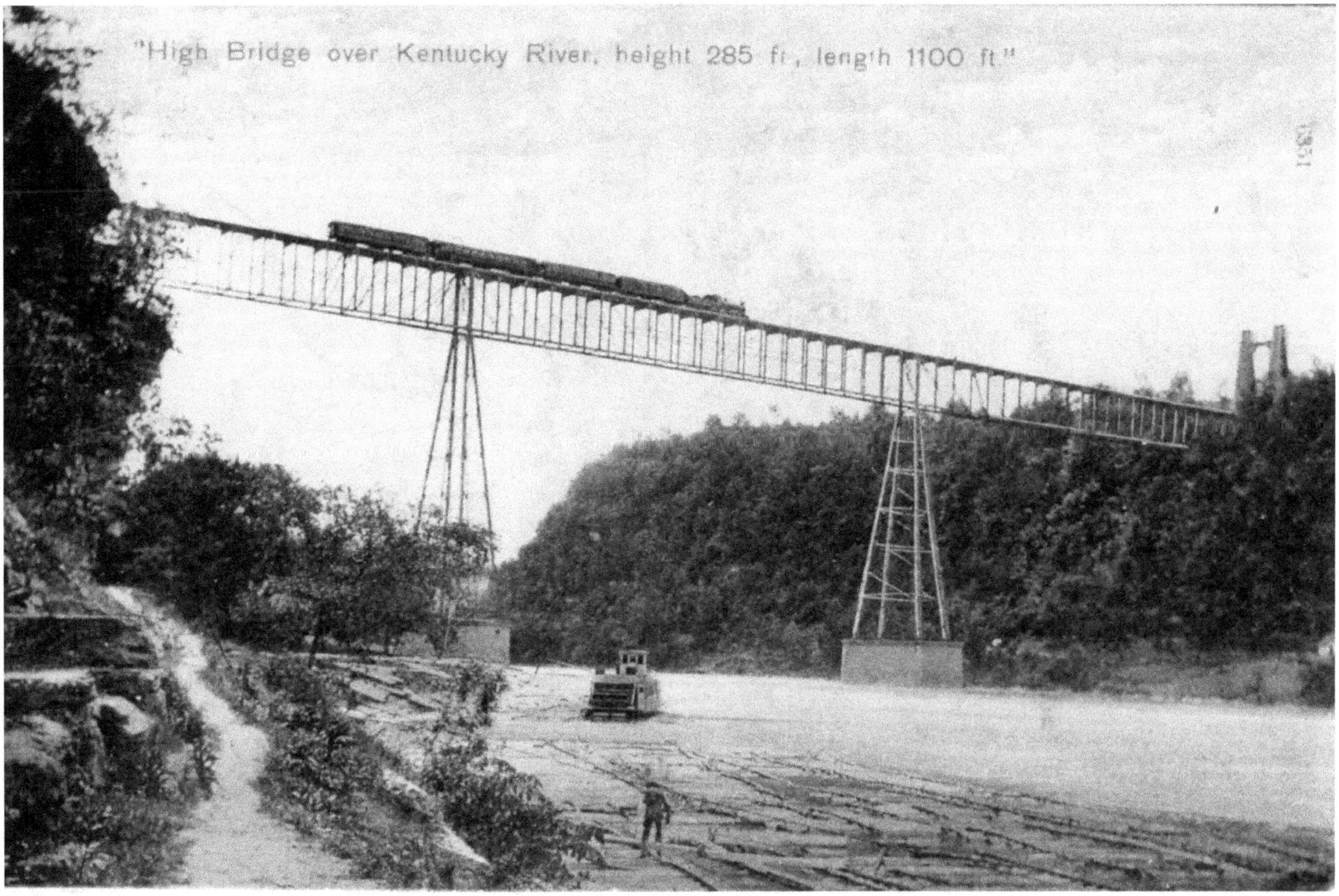

"High Bridge over Kentucky River, height 285 ft., length 1100 ft."

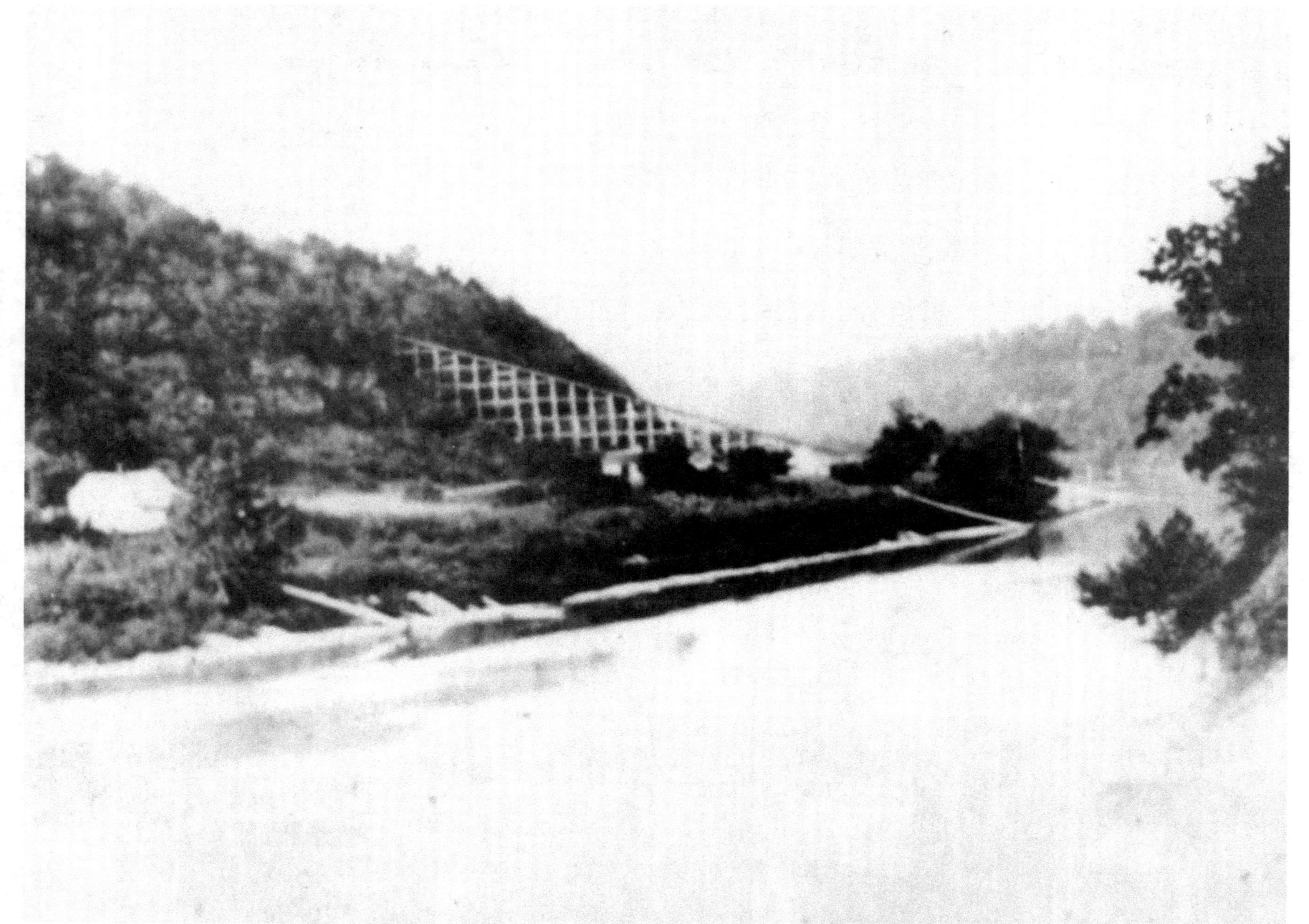

Because of Lock and Dam No. 7, High Bridge Lumber moved to land atop the cliffs. A log tramway was constructed to carry the logs from the river up to the sawmill. Area native Ken Houp recalls a tale his grandmother told him of finding an unbranded log as a little girl. Not wanting to lose it to someone, she rode the log up the tramway to the sawmill where she sold it.

Perhaps the only one still in existence today, this receipt from the High Bridge Lumber Company to the Danville Fishing Club is dated October 13, 1892, wherein the club purchased four pairs of oars made by the company for a total cost of $4.00.

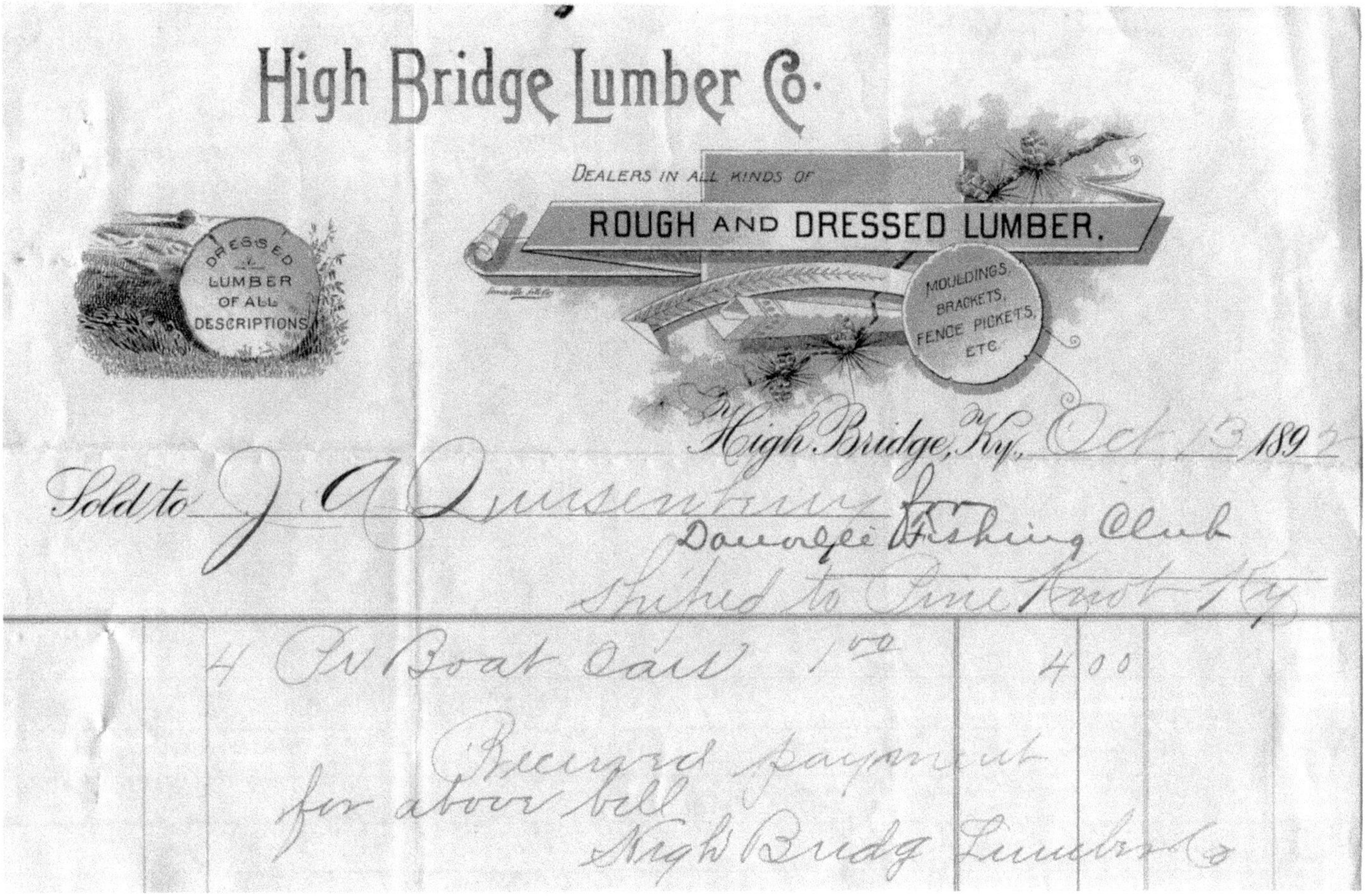

High Bridge Lumber Co.

Dealers in all kinds of

ROUGH AND DRESSED LUMBER.

Dressed Lumber of all Descriptions

Mouldings, Brackets, Fence Pickets, Etc.

High Bridge, Ky., Oct 13 1892

Sold to J. A. Quisenberry for Danville Fishing Club

Shiped to Pine Knot Ky

4 Pr Boat Oars	1 00	4 00	

Received payment for above bill

High Bridge Lumber Co

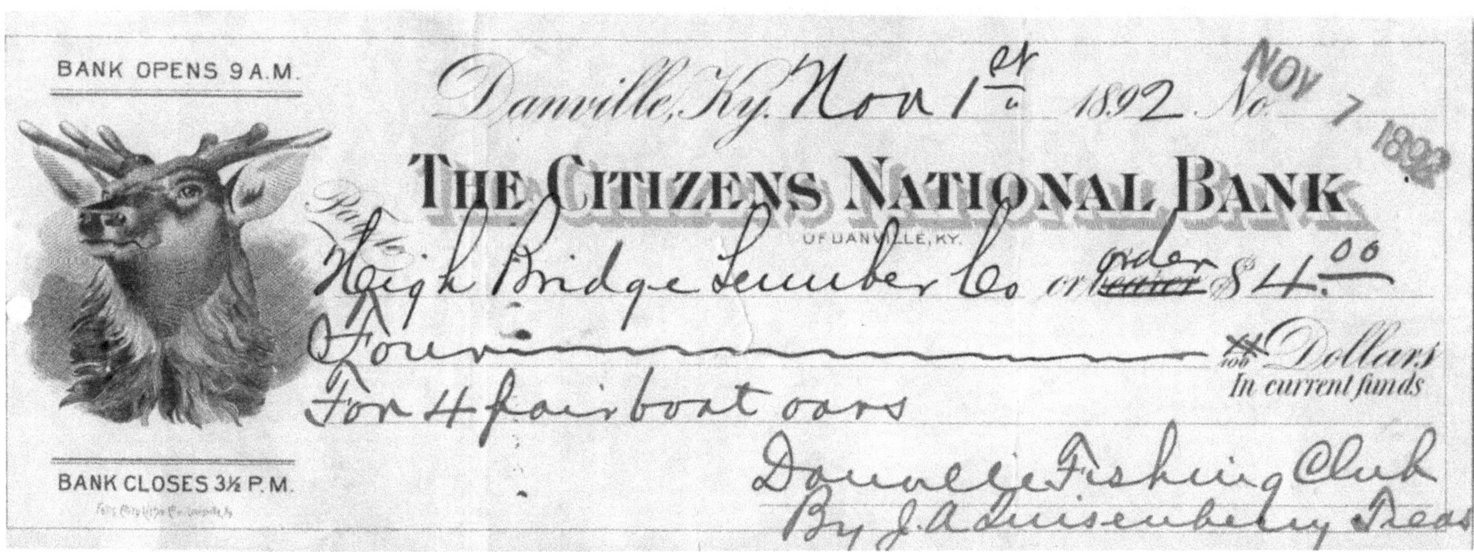

BANK OPENS 9 A.M.

Danville, Ky. Nov 1st 1892 No. 7

NOV 7 1892

THE CITIZENS NATIONAL BANK OF DANVILLE, KY.

Pay to High Bridge Lumber Co or order $4.00

Four xx/100 Dollars In current funds

For 4 pair boat oars

Danville Fishing Club

By J. A. Quisenberry Treas

BANK CLOSES 3½ P.M.

As a companion to the receipt, this is the check written out by the Danville Fishing Club to the High Bridge Lumber Company for the purchase of the oars. The draft was payable on the Citizens National Bank, signed by J.A. Quisenberry, and the check was presented for payment at the Citizens National Bank in Danville on November

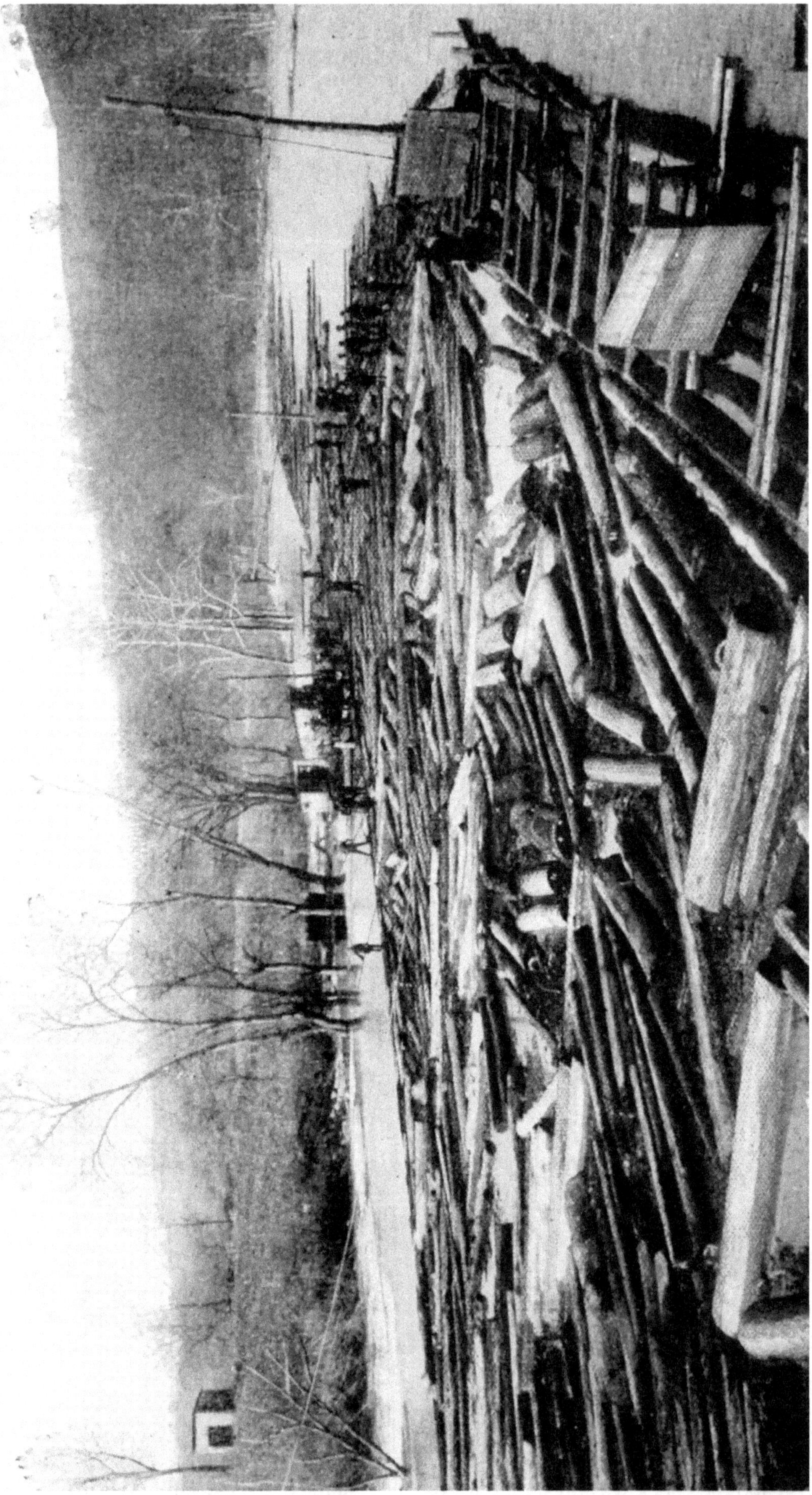

Logging on Kentucky River, Kentucky.

To reach the river below, visitors to High Bridge Park originally had to travel a distance outside the park and descend a cliff side road. Wanting to make it easier for patrons to reach the river bottom to swim, fish, and marvel at the height of the bridge below, in May 1902 a plan was devised to build stairs from the park down the side of the cliff.

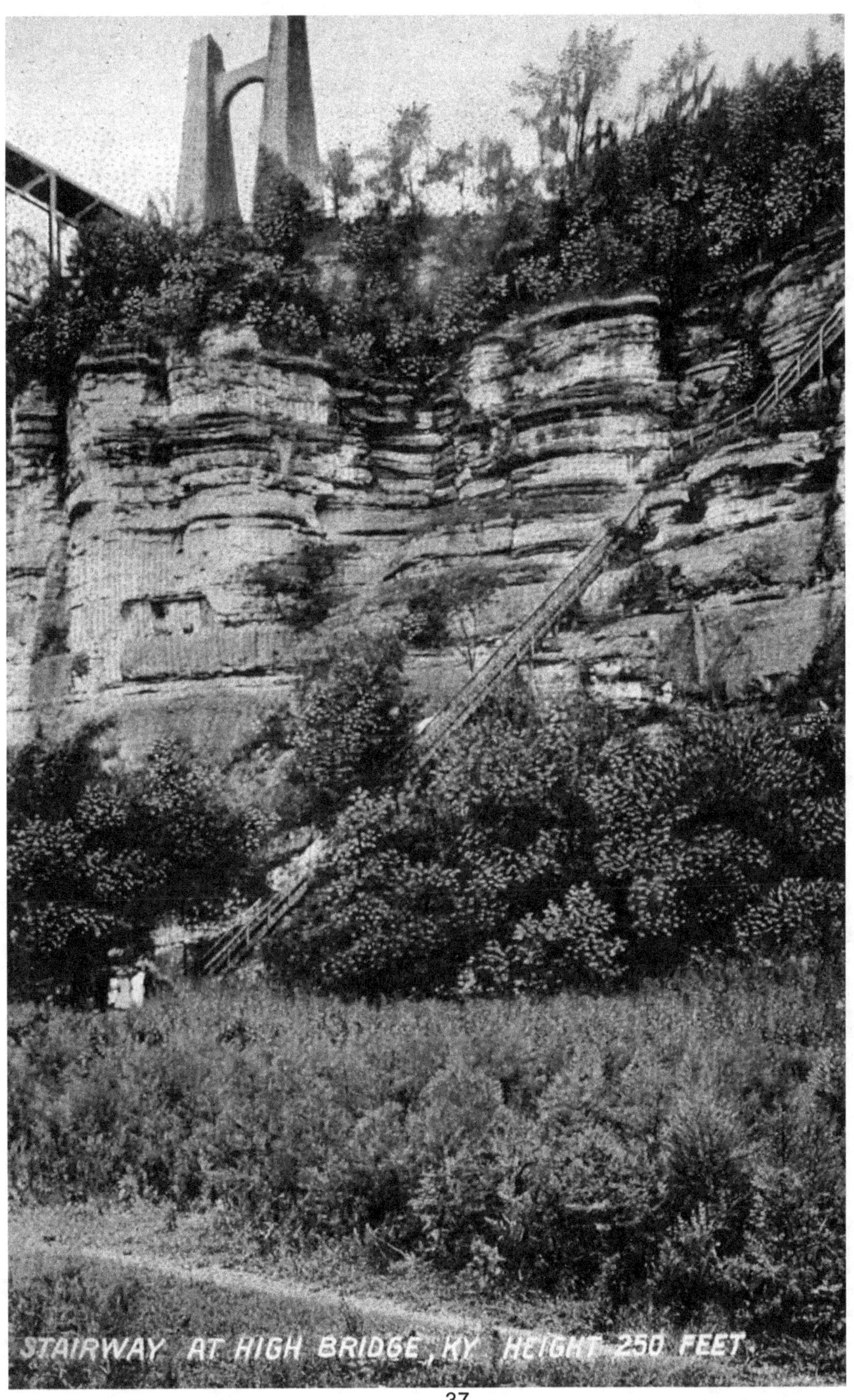
STAIRWAY AT HIGH BRIDGE, KY. HEIGHT 250 FEET.

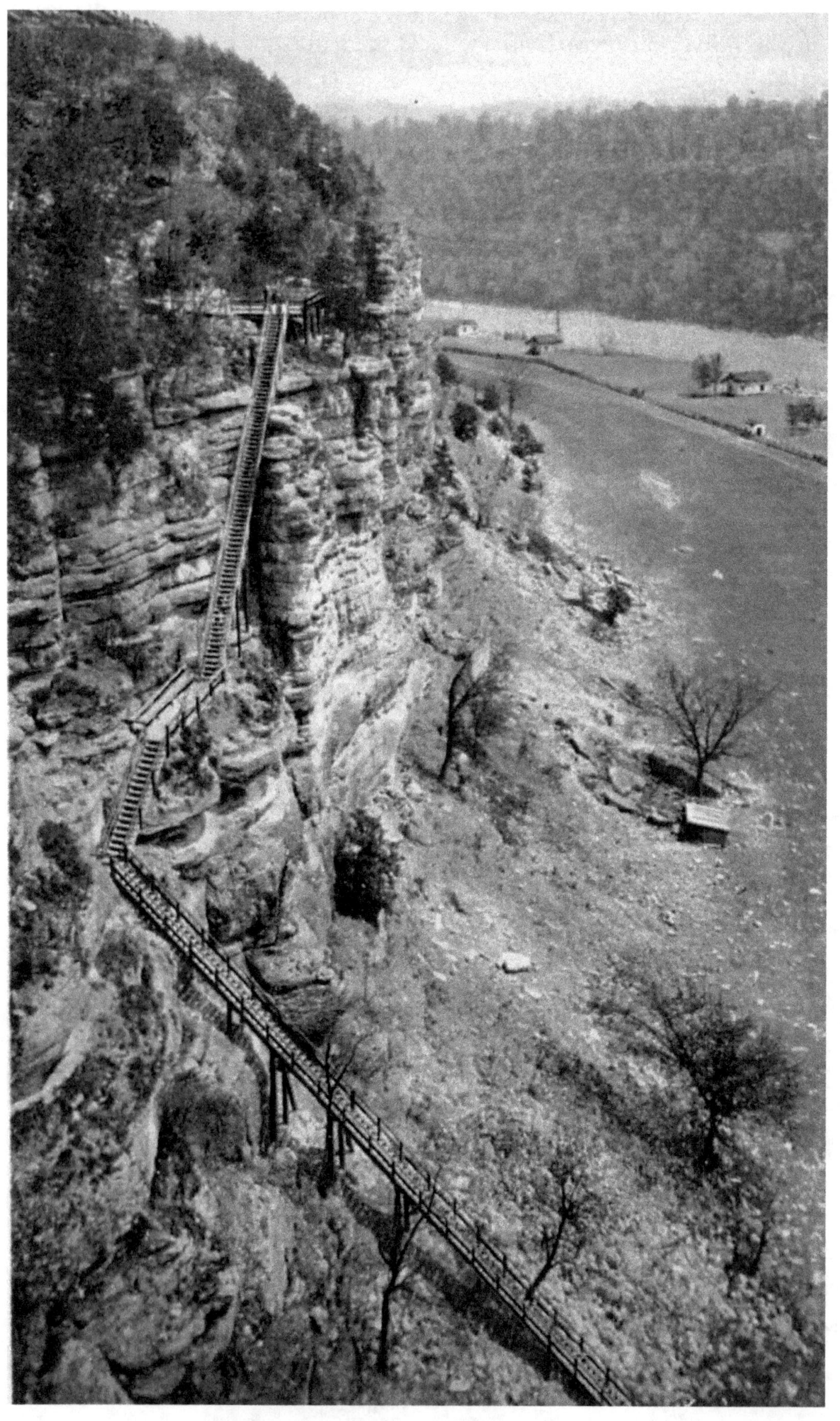

High Bridge resident Professor Barstow led the movement to build the stairs down the face of the cliff. The staircase was presumably made of lumber from the High Bridge Lumber Company. Obviously for brave souls, the staircase consisted of 271 stairs and was completed in August 1902. A woman was quoted as saying she went down the stairs but, "Once was enough!"

Throughout its life, High Bridge has been crossed by regular folks and notables alike. One such notable person was Prince Henry of Prussia, who crossed the bridge on his way to visit Louisville in March 1902. Upon word of the impending royal crossing, many residents turned out to catch a glimpse of his train as it passed.

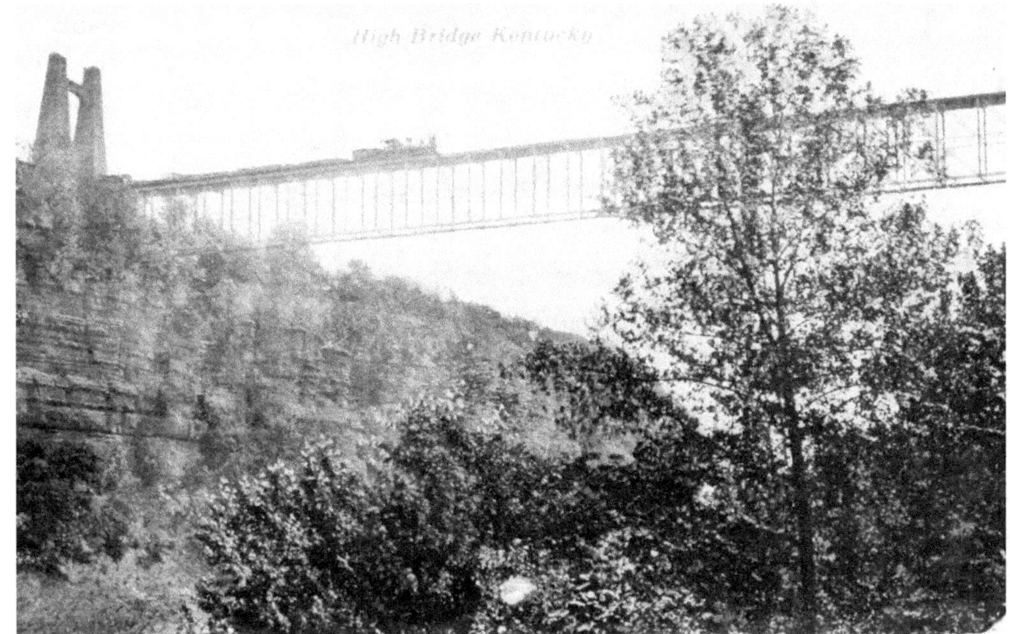

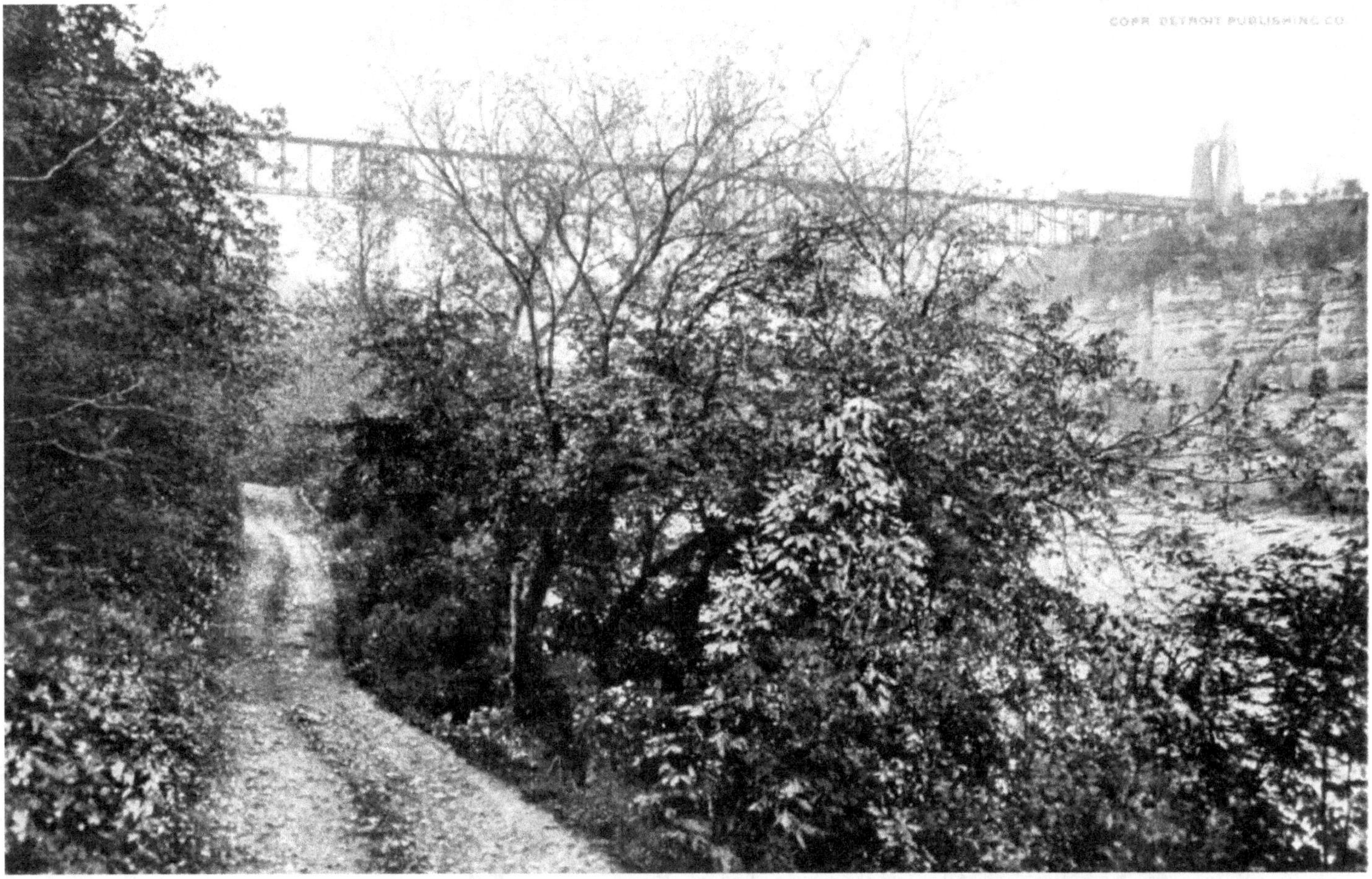

COPR DETROIT PUBLISHING CO

The Old Wilderness Road followed the riverbank on the Mercer County side of the river, and it also led to the Shaker Ferry Landing. Along the road many cabins could be found. The road was also an ideal setting for visitors to the area to hike along the river and enjoy the view of the towering palisades and the bridge.

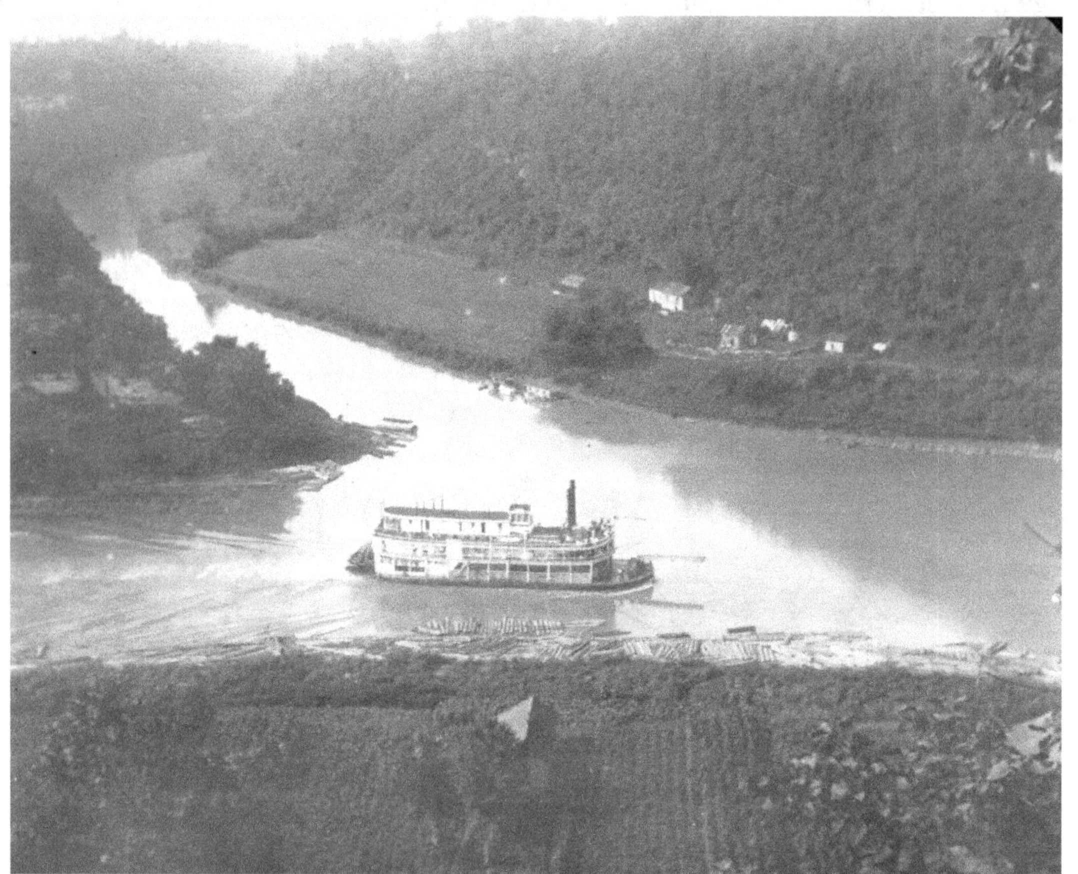

As word spread of the beauty of the High Bridge area, riverboat excursions became very popular. The *Falls City II*, already a popular river excursion vessel, made weekly trips to High Bridge, sometimes bringing excursionists from as far away as Indianapolis, Indiana, and Benton Harbor, Michigan, who had traveled to Louisville for the sole purpose of boarding the vessel for a trip to High Bridge.

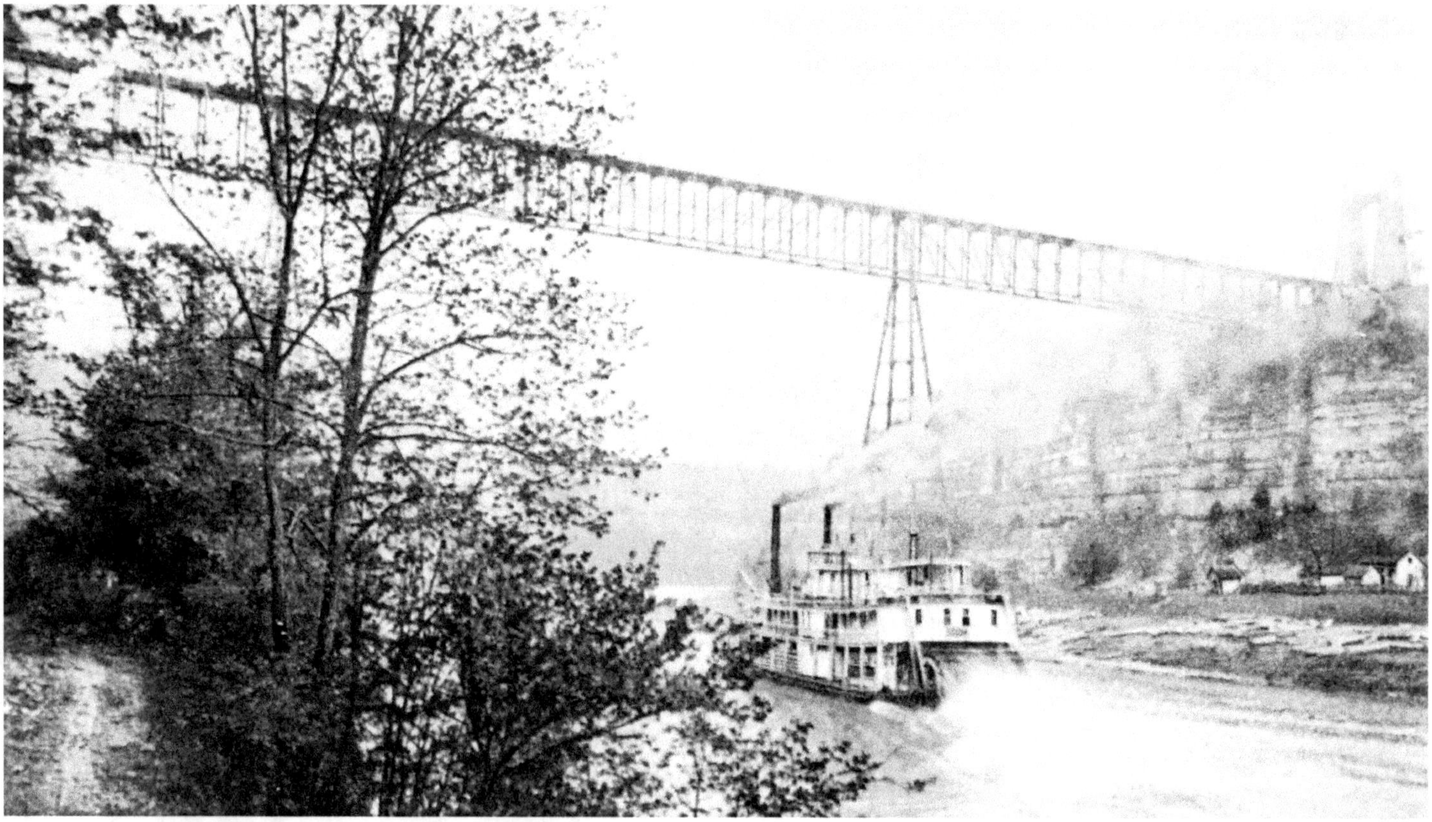

The familiar whistle blast of the *Falls City II* was a welcome sound to High Bridge residents. Due to the diligence of Capt. Squire Jordan Preston and Capt. Jonathan Newton Abraham, the sighting of the vessel could be predicted like clockwork. Even when the ship was not coming into port in the community, many in the area would run to the cliff edge or riverside to greet the vessel as it passed.

The river was the life of Capt. Squire Jordan "S. J." Preston, born in 1850. Through the years he served as one of the captains of the *Falls City II* and was known for his robust personality.

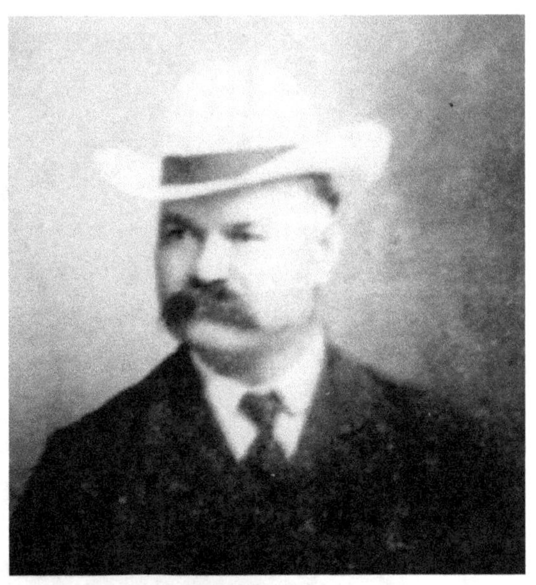

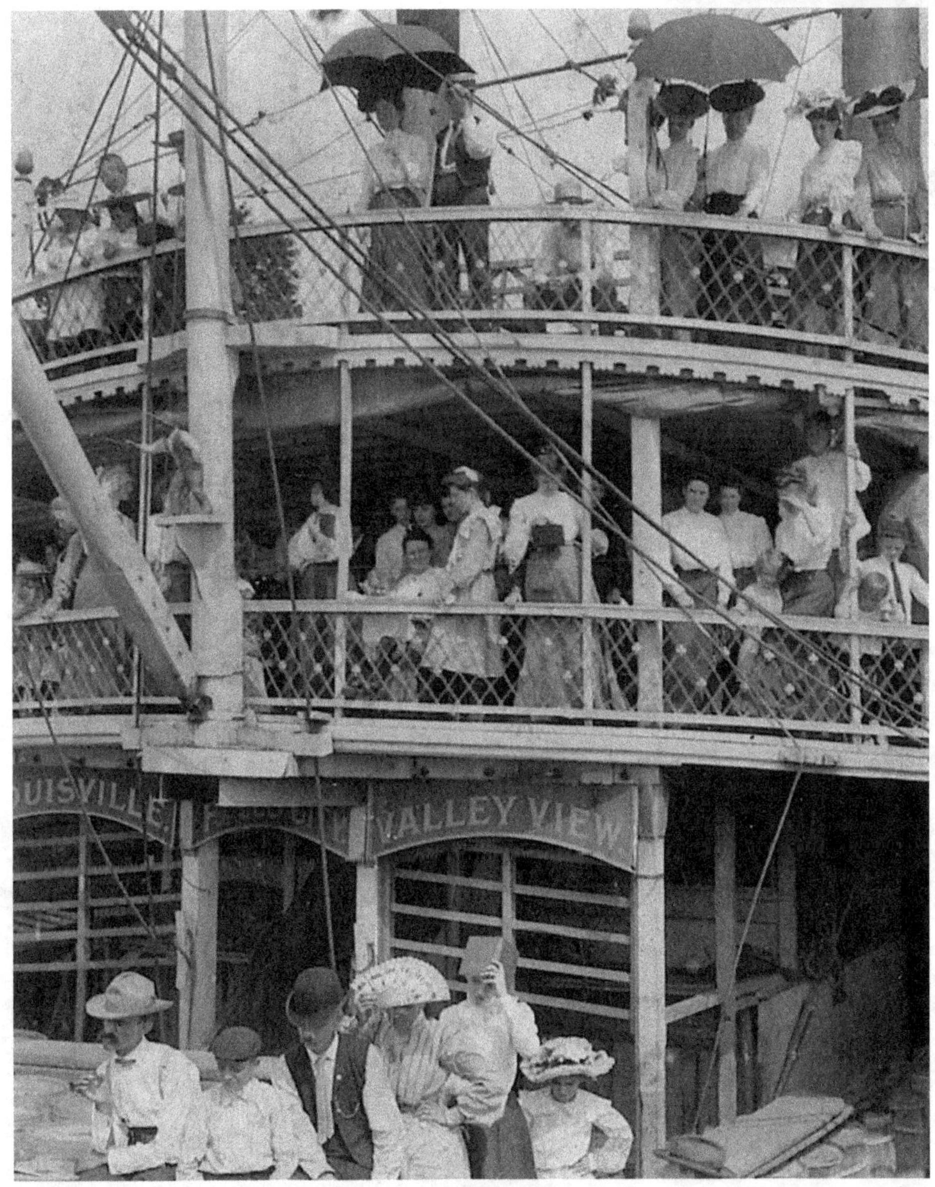

Although at times the conditions on board could be crowded, upon their return home excursionists would recount the fact that their minor discomforts were overlooked because the officers and crew were so "gentlemanly, clever and obliging." However, if you were late arriving at the landing, you were left behind because the ship's officers wanted to keep it as punctual as the train.

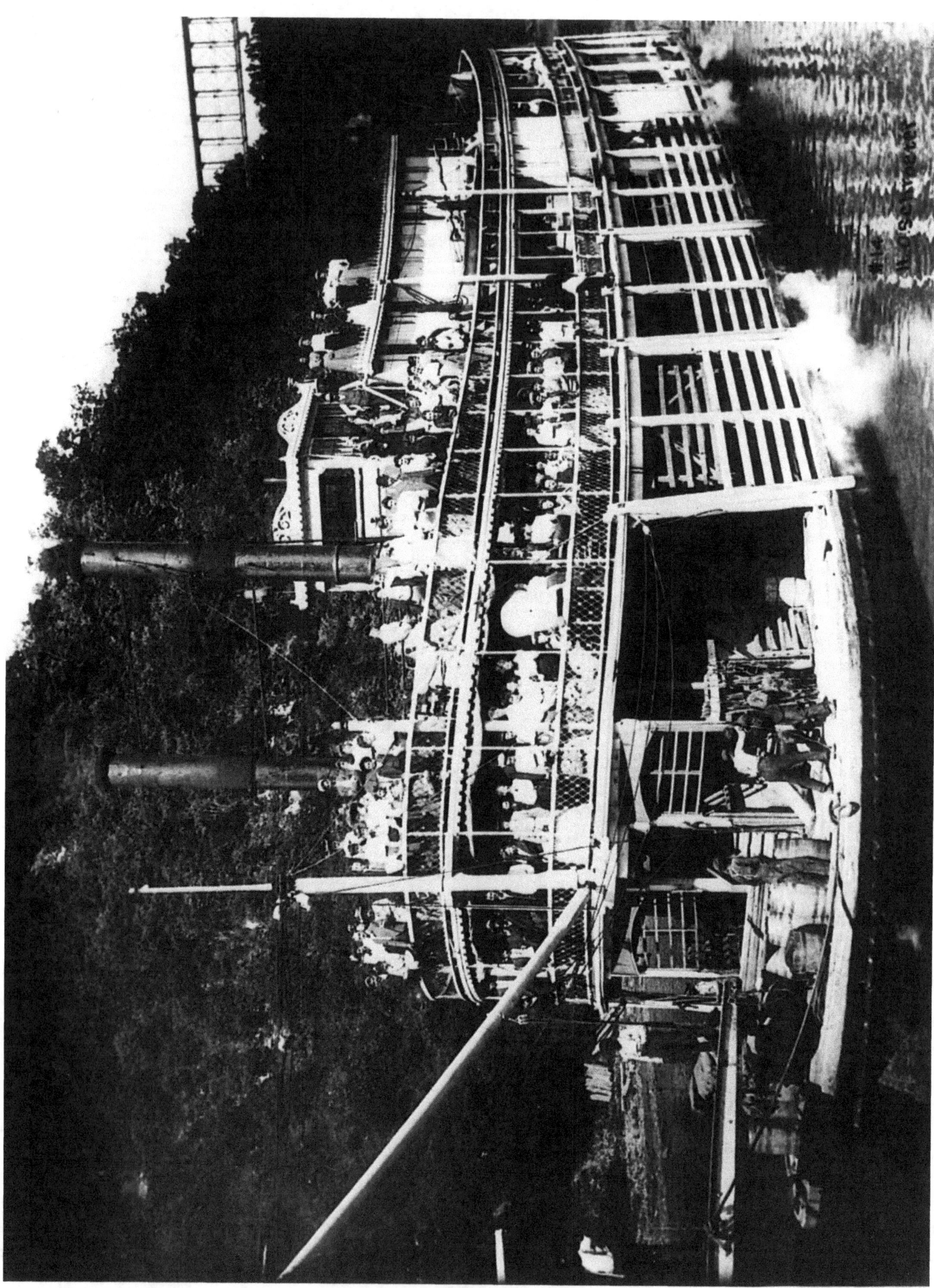
#14
W. O. Schroeder

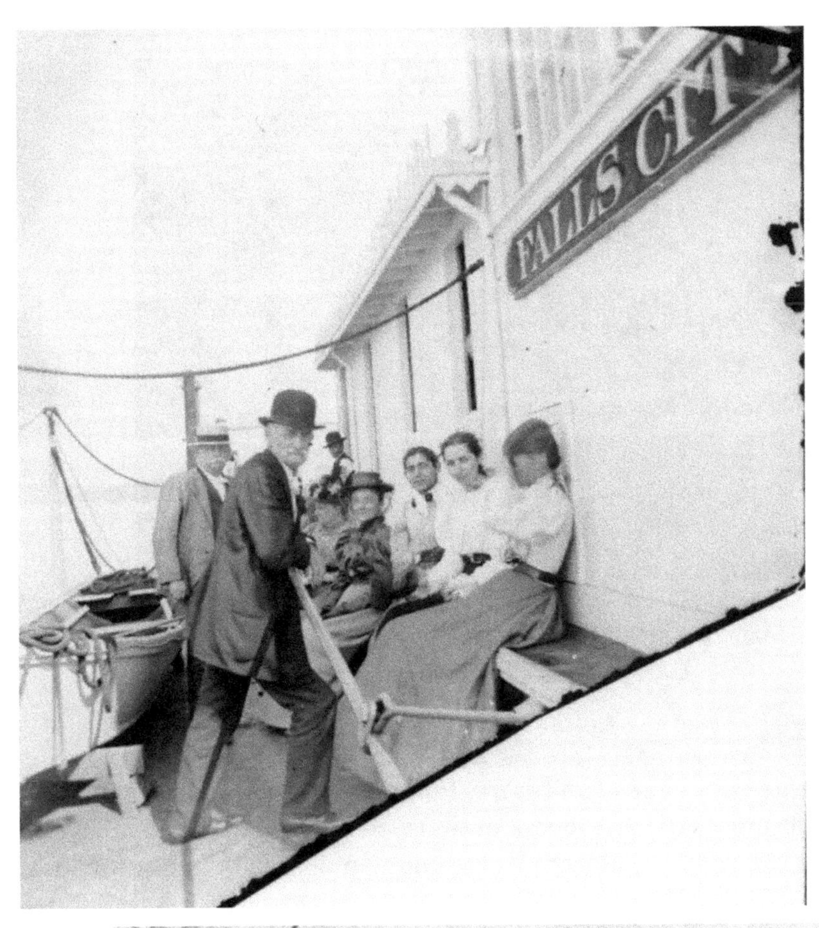

These families were among the many people to enjoy excursions to High Bridge on board the *Falls City II*. Excursions to High Bridge would often coincide with events held at High Bridge Park, such as the noted aquatic performer Captain Blondell. His show reenacted noted battles with model ships. All in attendance were promised a scene that was "realistic in the extreme" with his "costly models."

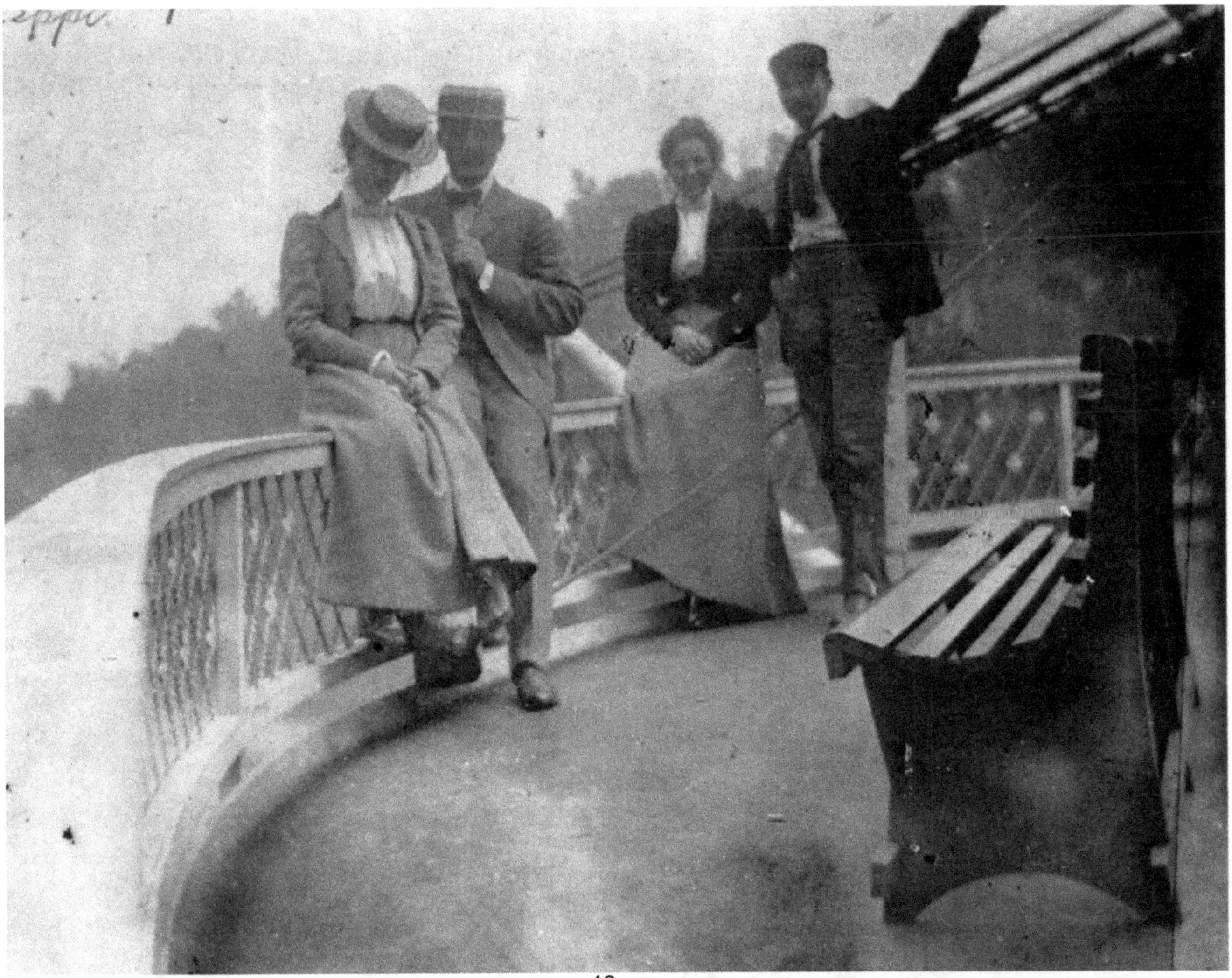

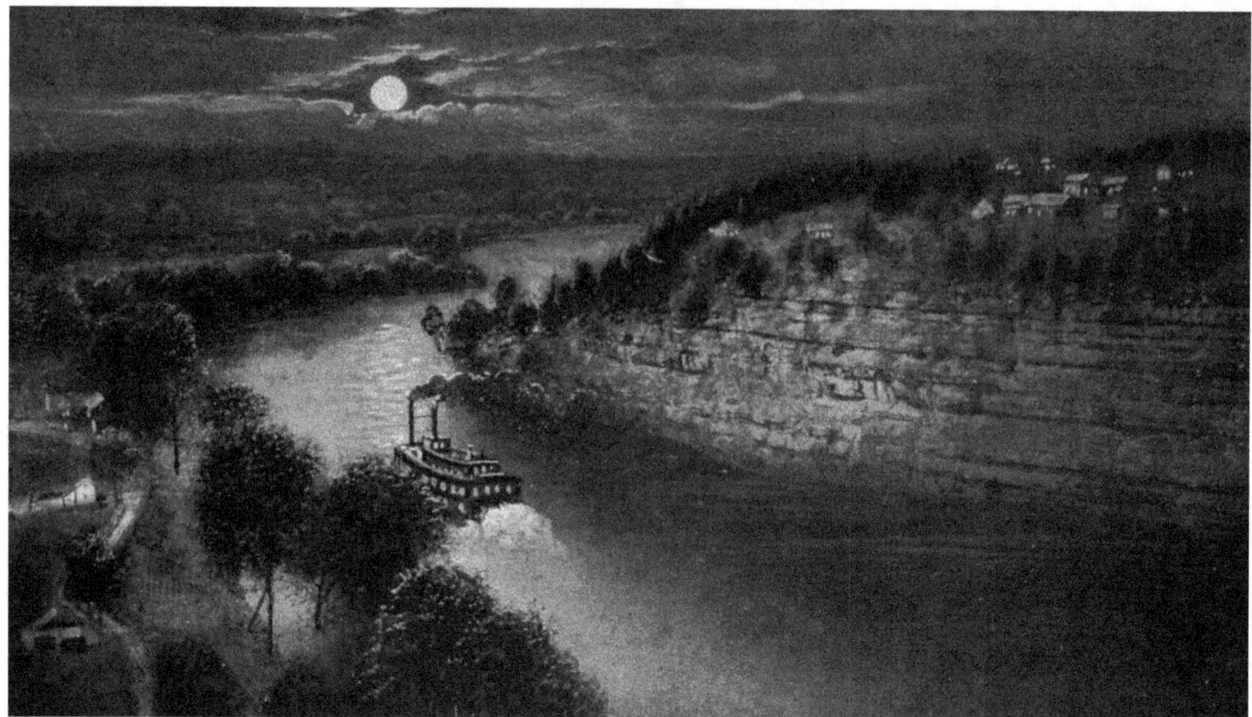

Night cruises were an alternative to the usual daytime excursions run by the *Falls City II*. In advertisements for such cruises, potential passengers were lured by promises of "an elegant outing free from all the annoyances of an excursion train, as well as the cool breezes of the river." The advertisement also claimed, "these torrid days require cool breezes in the evening to make good sleep at night."

The below photograph of the Pepper family was taken on a night cruise on the *Falls City II*. The excursions ran every few evenings and offered low rates to make it within the reach of all. The vessel furnished everything that was needed for the comfort of passengers on the excursions, which added to the experience and undoubtedly made it worth every cent paid.

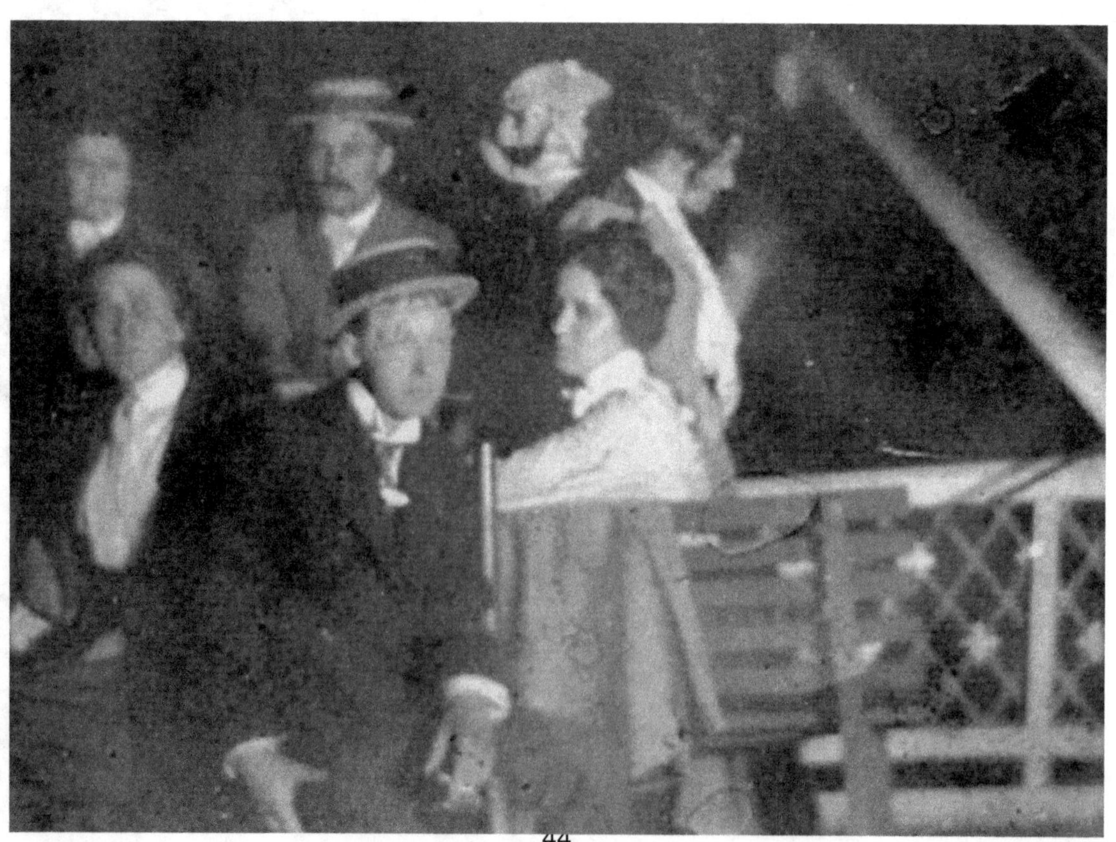

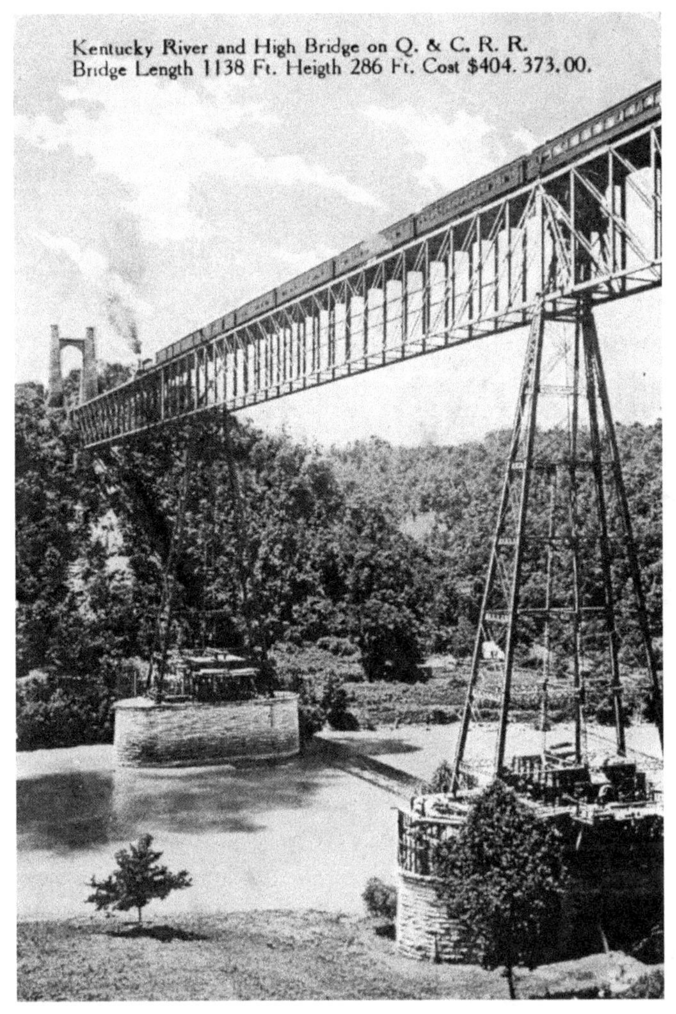

Tourists could also visit High Bridge by train. The Queen and Crescent Railroad and later the Southern Railroad heavily promoted train excursions from Lexington and Cincinnati to High Bridge. It was often billed as a "delightful place for boating, bathing, fishing and bicycling." Special fares were offered for these excursions.

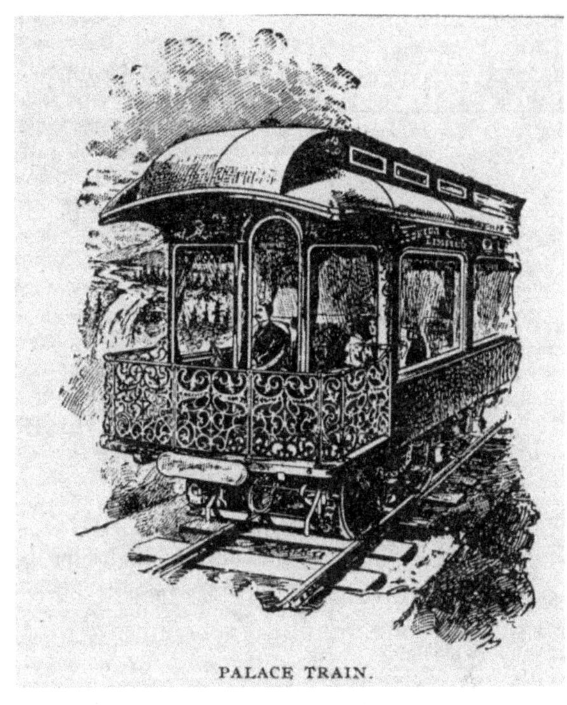

PALACE TRAIN.

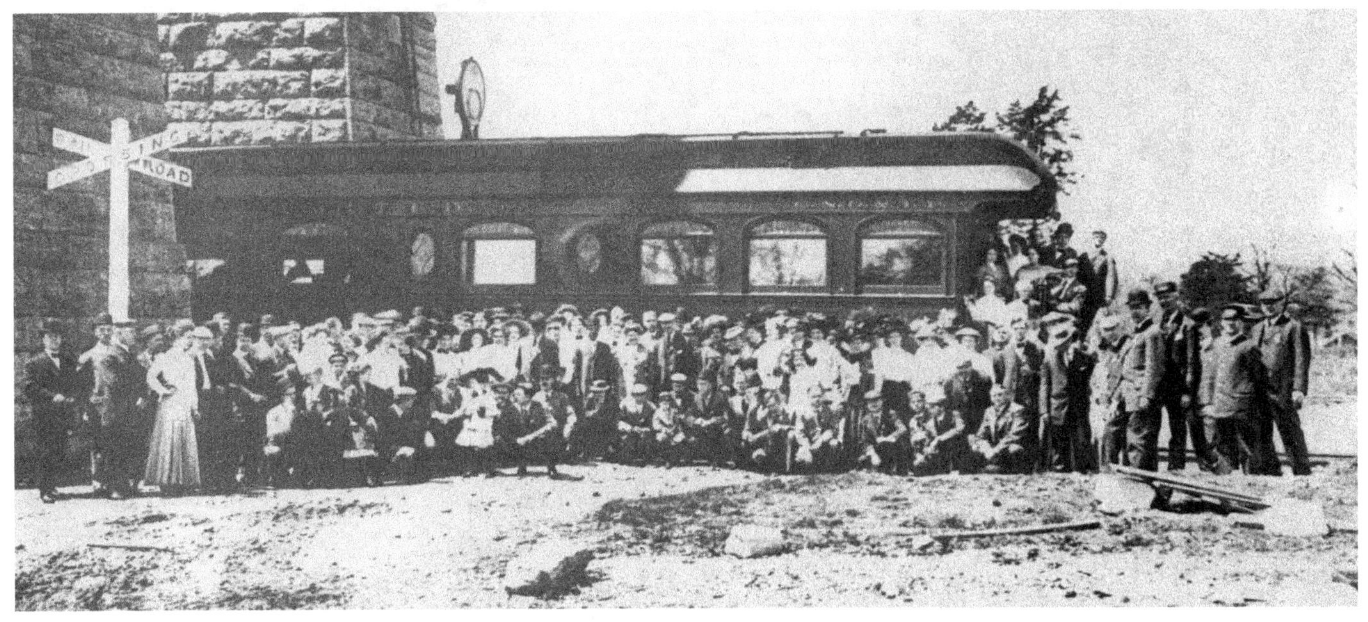

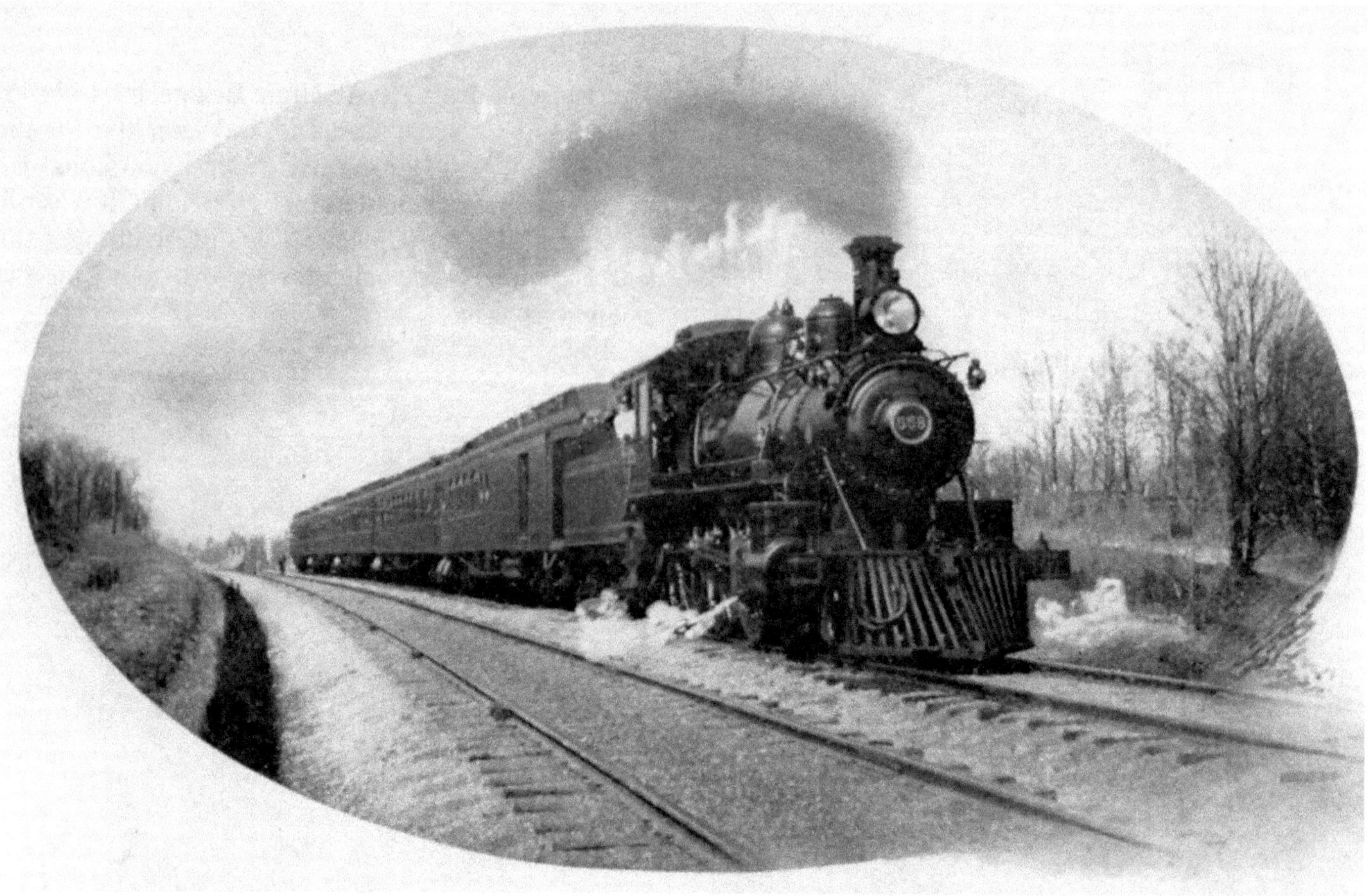

A *Kentucky Gazette* advertisement dated July 5, 1899, read as follows: "Excursions are run from Cincinnati and all points south every other Sunday to High Bridge. There isn't a more beautiful place to visit than this entrancing spot. You can't afford to miss it. $1 from Cincinnati, 50¢ from Lexington and like low rates from other points. Ask your ticket agent about it. You cannot afford to miss this pleasant day on the river."

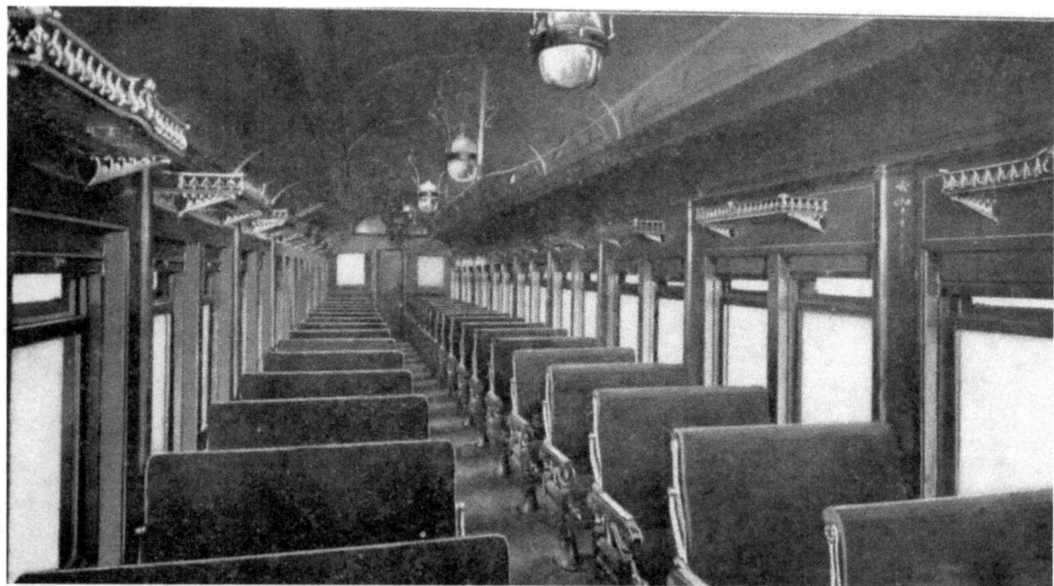

The Queen and Crescent Railway boasted about their train cars, "Their interior is as fine as can be constructed, and of the best material while from the outside they are dreams of beauty."

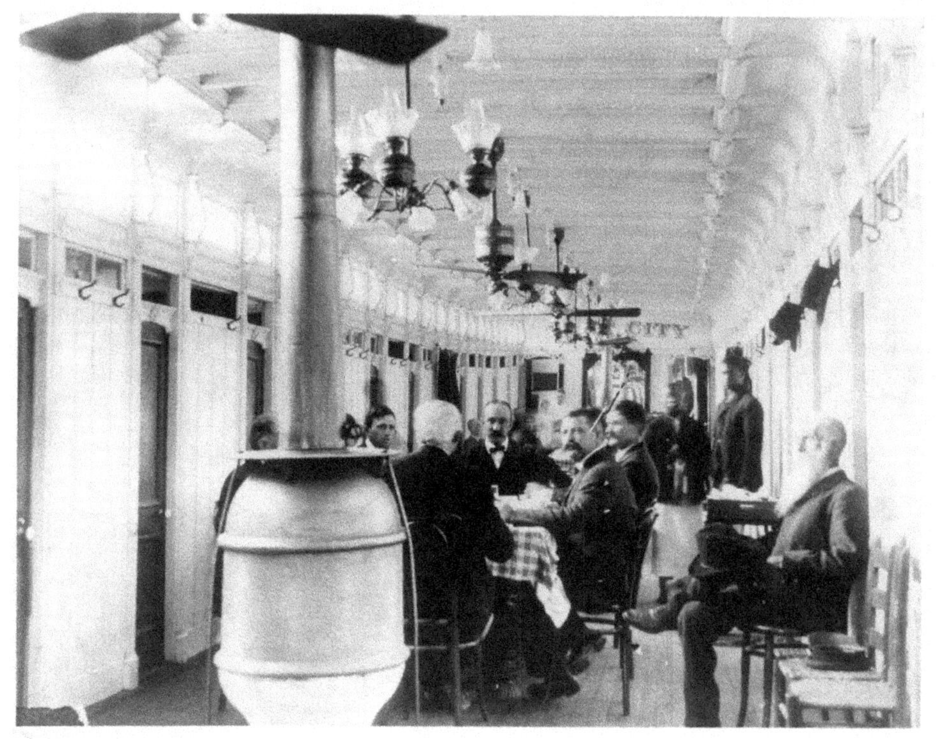

Circle Excursions were operated by the Queen and Crescent Railway in conjunction with the *Falls City II*, with fares ranging from $1 to $1.25. One could take the train from Frankfort to High Bridge, spend time in the park, board the *Falls City II*, and enjoy dinner, music, and dancing on a night cruise back to Frankfort, where one would disembark the vessel to one's respective train to return home.

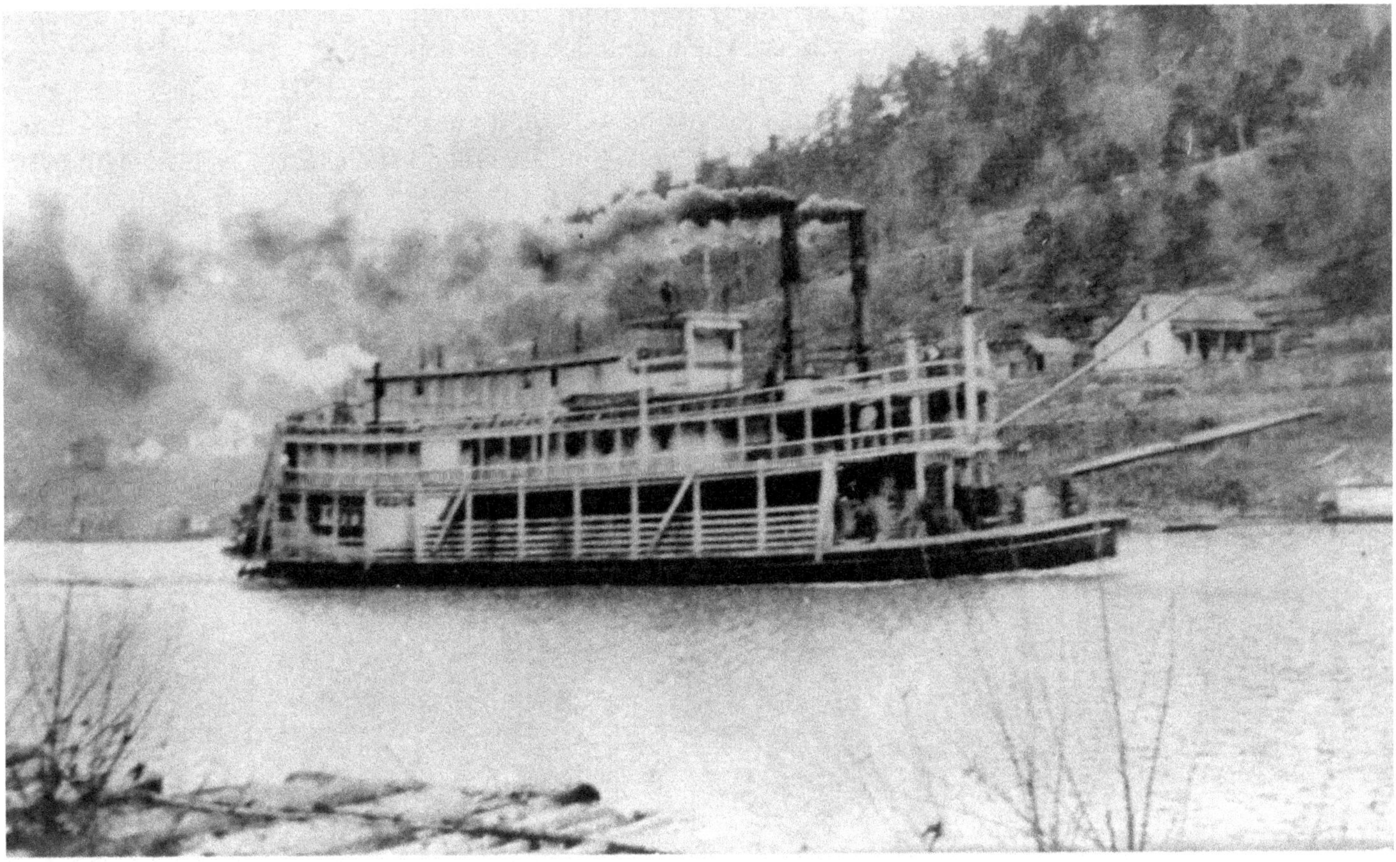

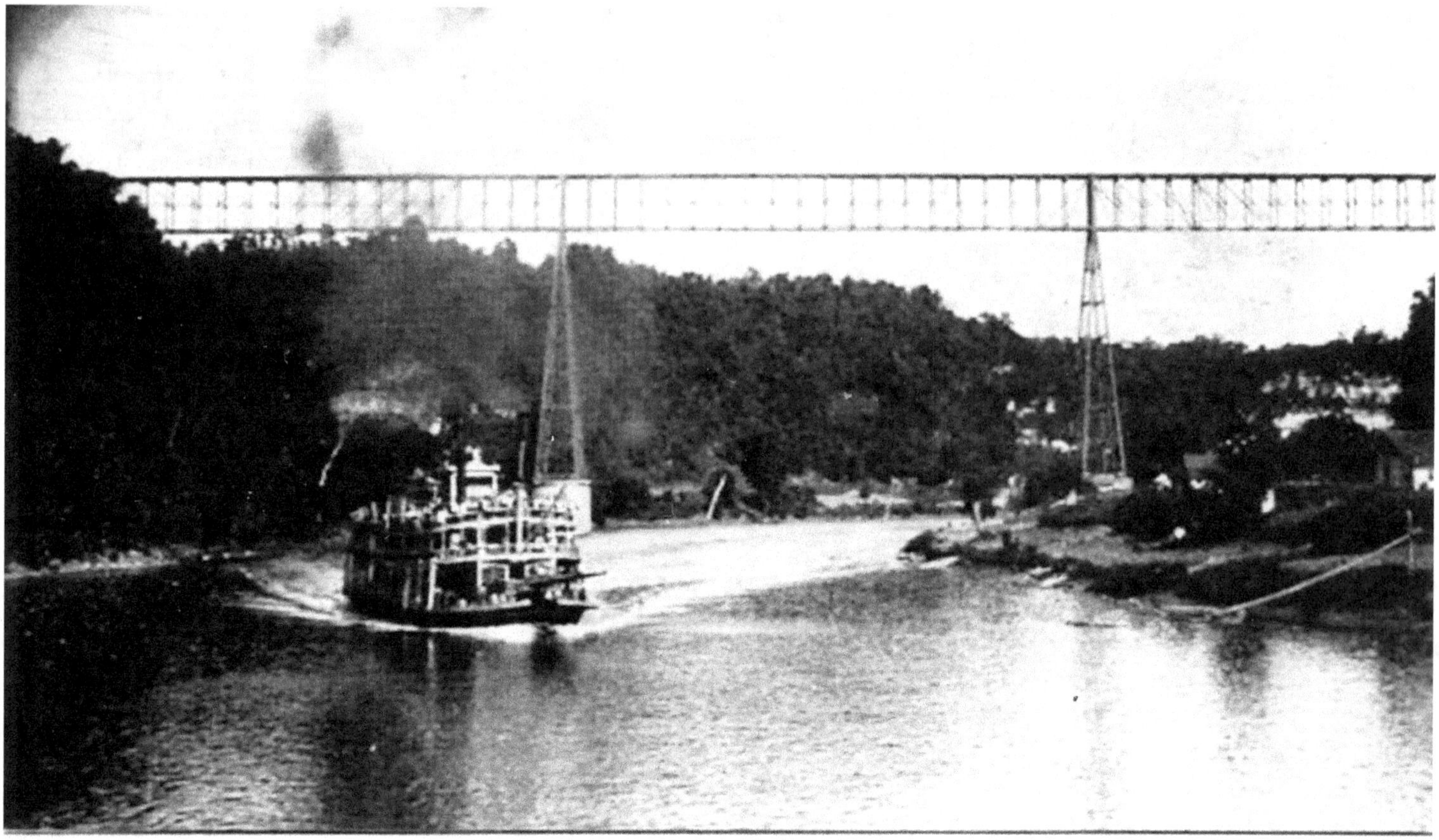

Kentucky River Excursion Co. Steamer Corker with a crowd of tourists passing under High Bridge. Height of bridge 326 feet above water.

The *Falls City II* was not the only vessel that brought excursionists to High Bridge. The Kentucky River Excursion Company ran the *Corker* from Louisville to High Bridge on Mondays during the summer months on a six-day voyage. The fare was $25 per passenger, which included meals and berth.

An advertisement postcard for the *Corker* invited the public to "spend your vacation with us on the Kentucky River, the Hudson of the West." It stated that the vessel was "safe and sound and as large as locks will permit, navigated by competent and courteous officers. Good clean berths [and] the best the market affords to eat. Reservations, accompanied by fare, must be made at least fifteen days in advance."

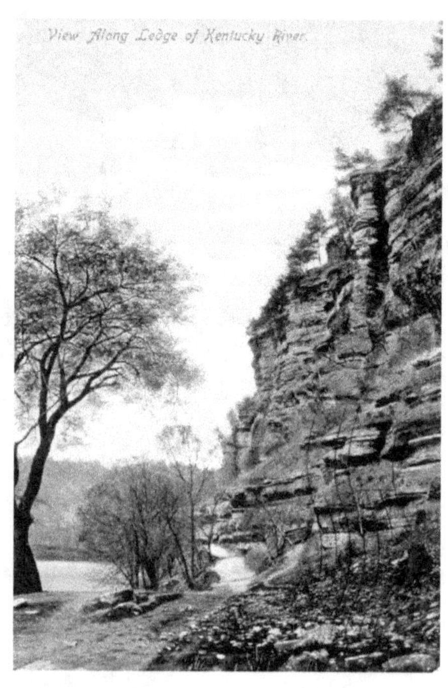

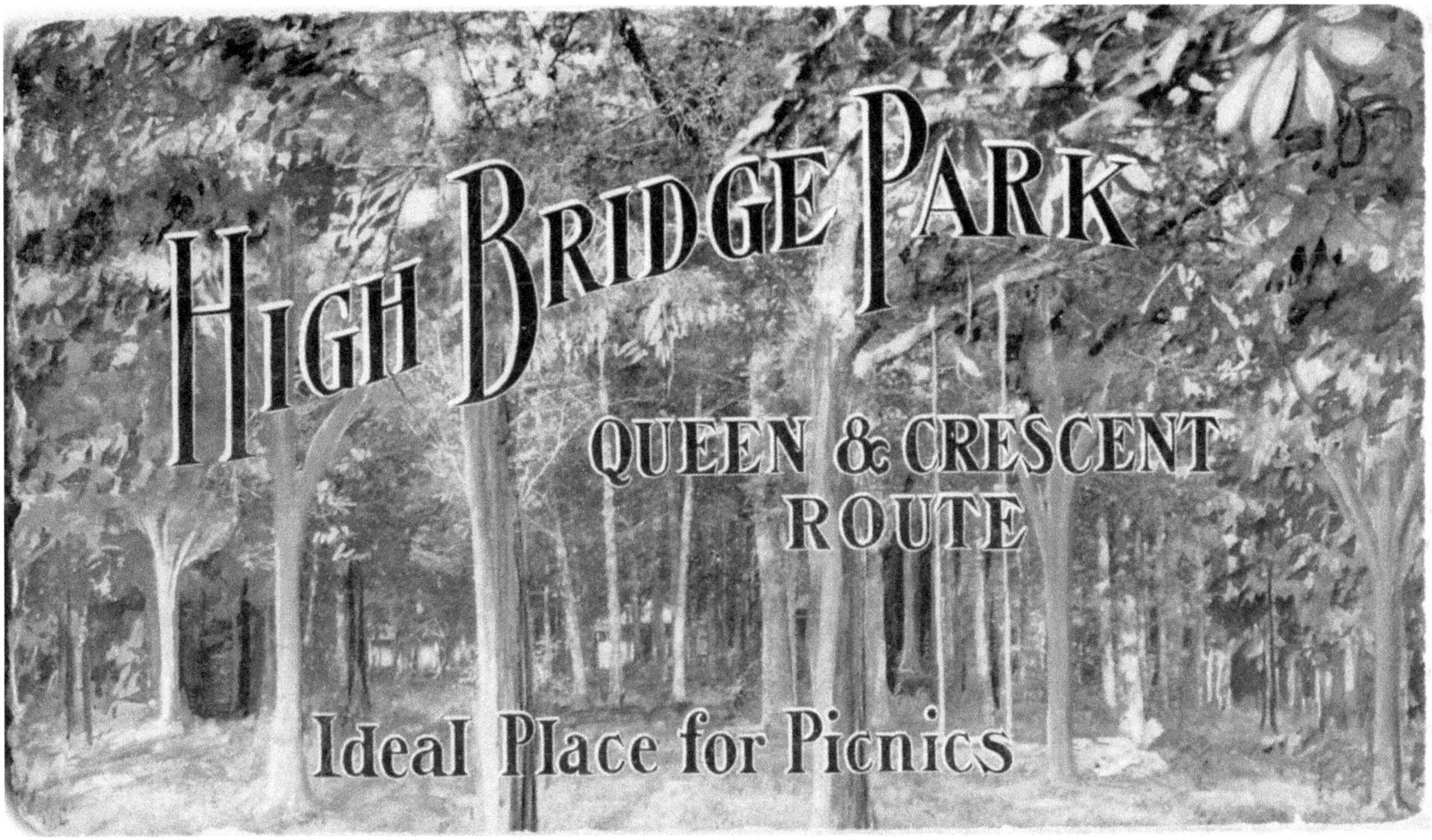

HIGH BRIDGE PARK

QUEEN & CRESCENT ROUTE

Ideal Place for Picnics

High Bridge was touted as having "The greatest picnic grounds in the country." by the Queen and Crescent Route.

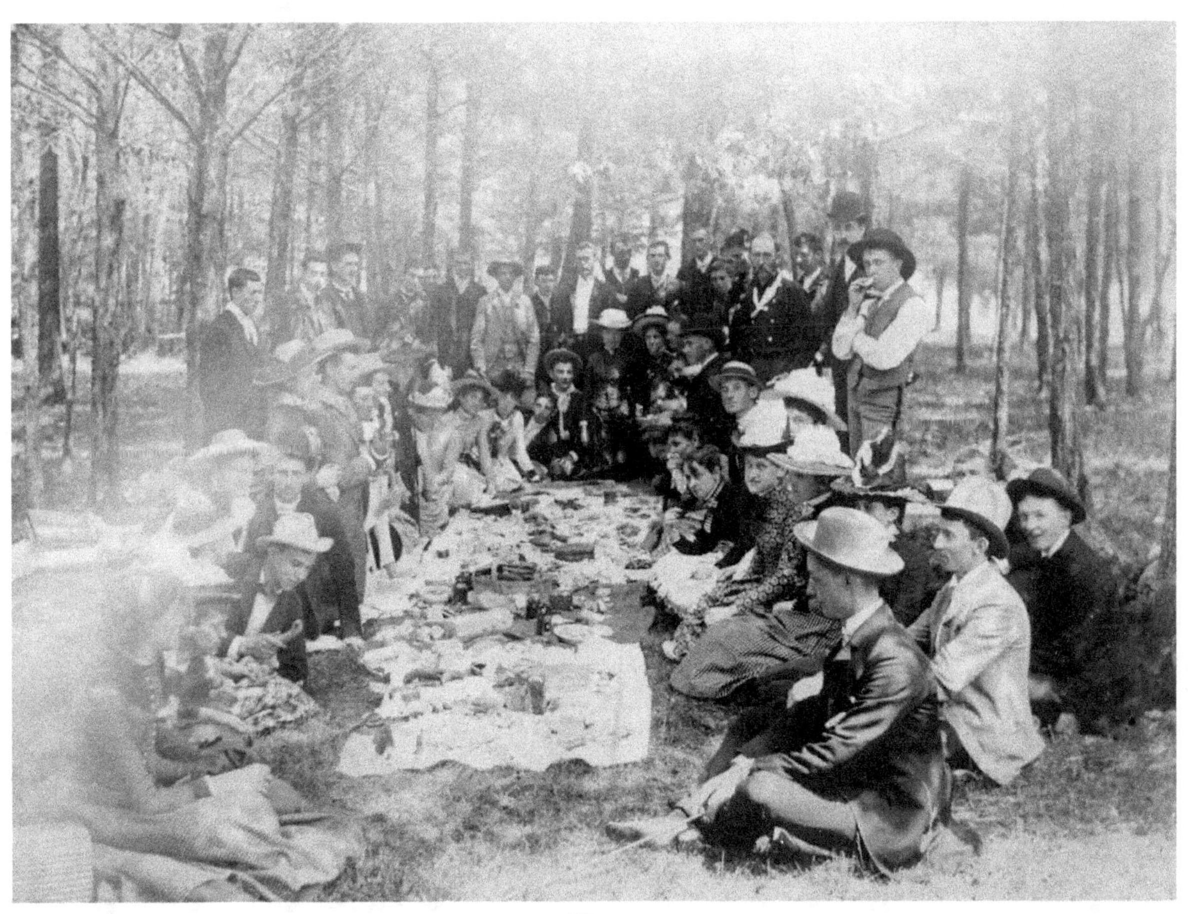

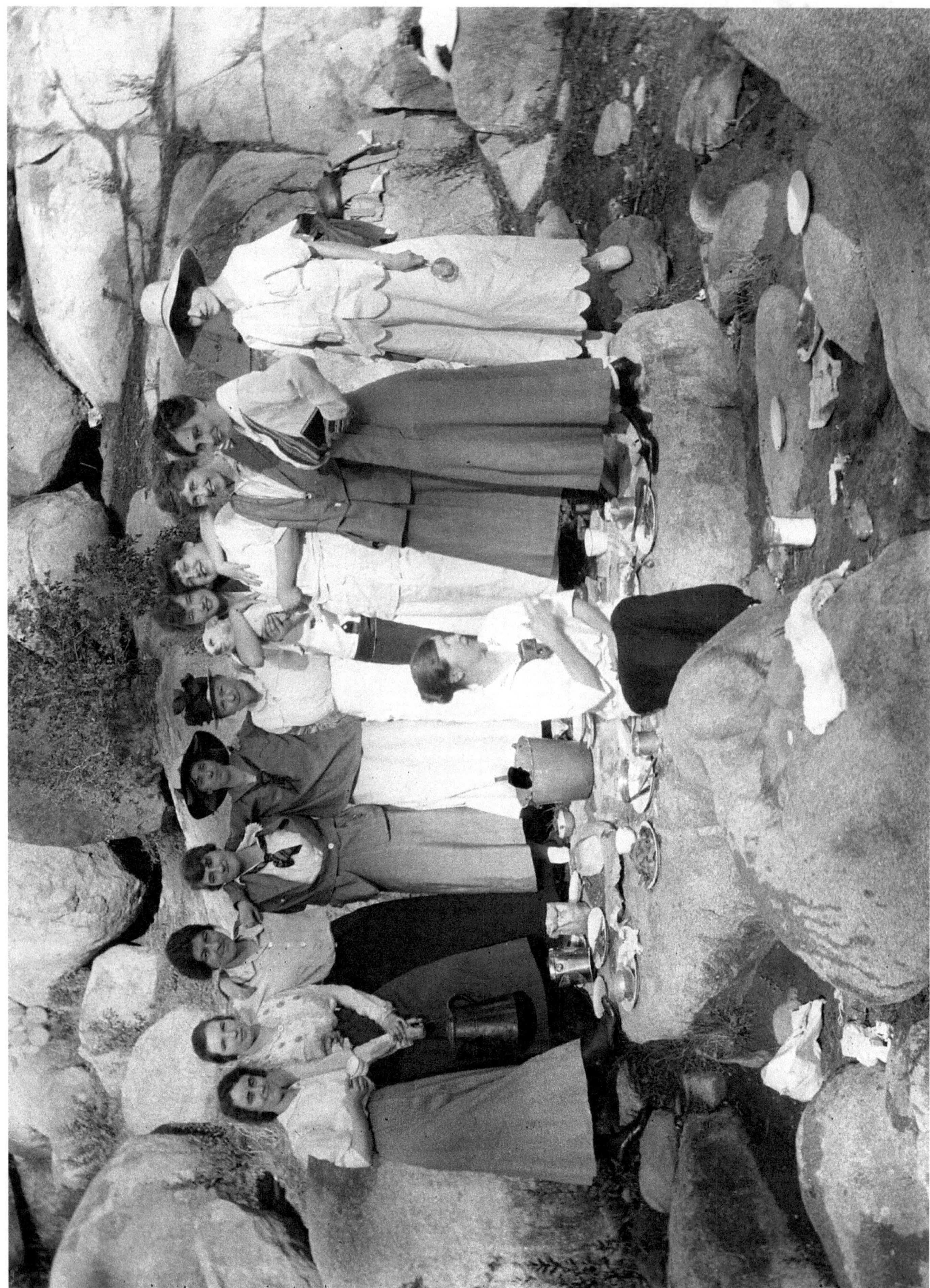

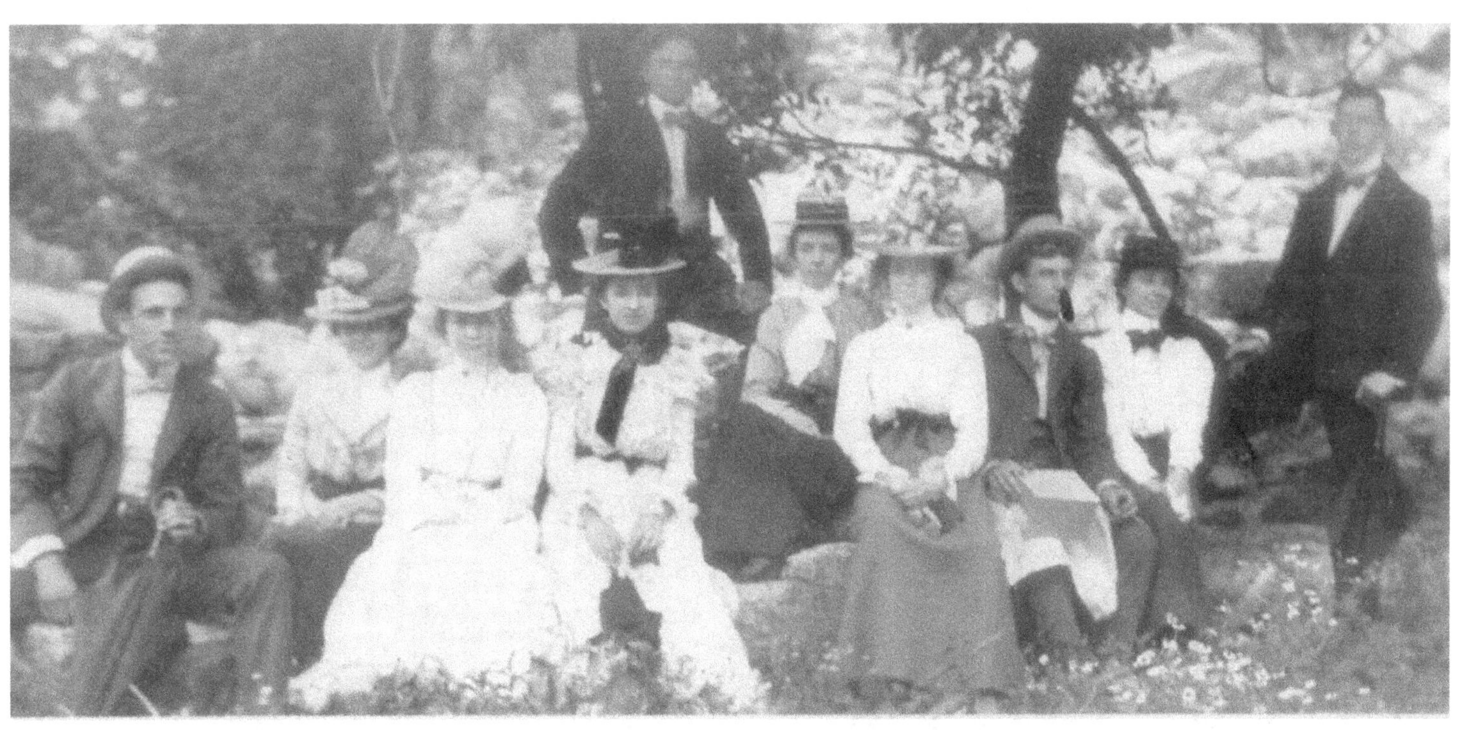

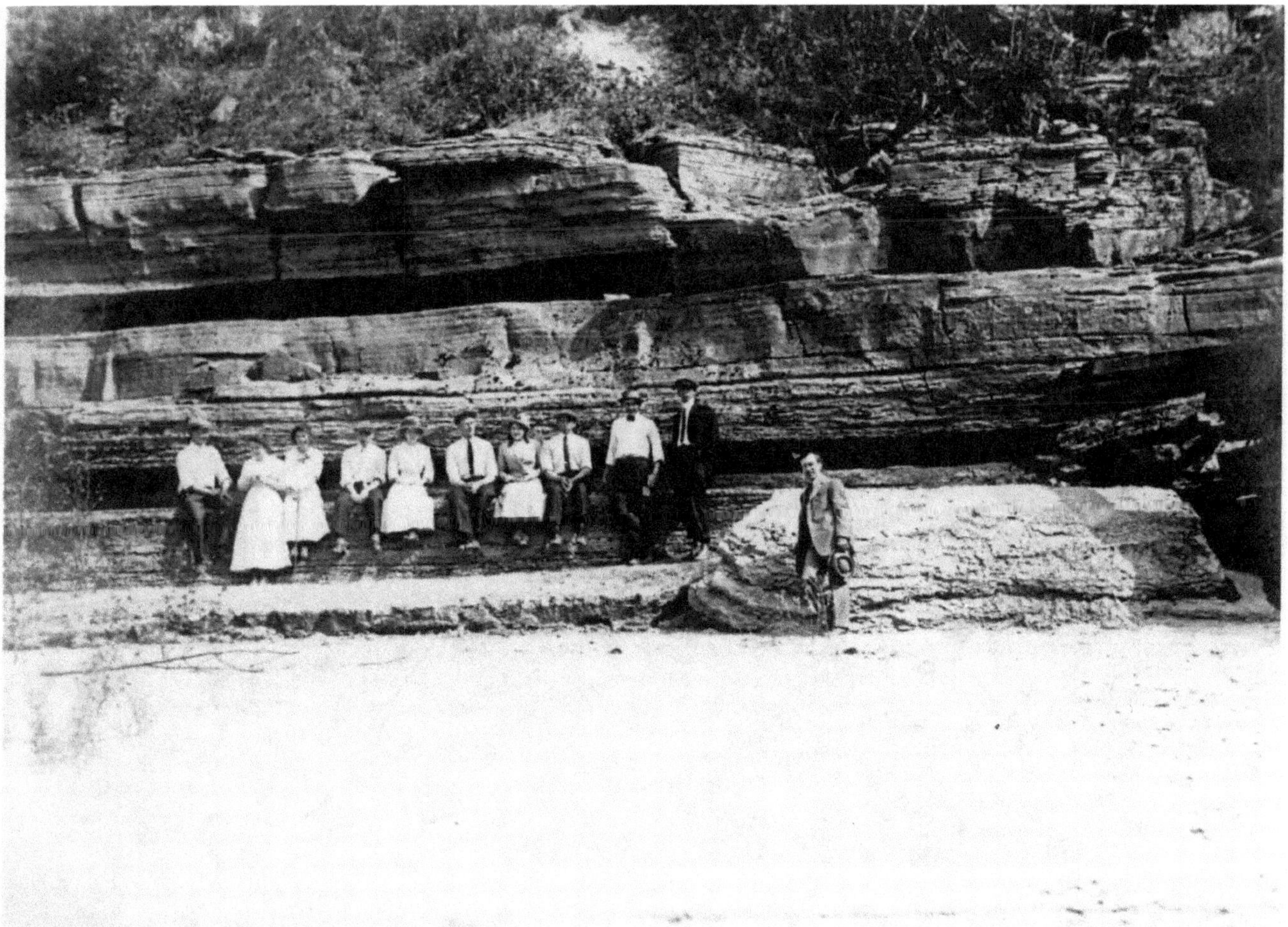

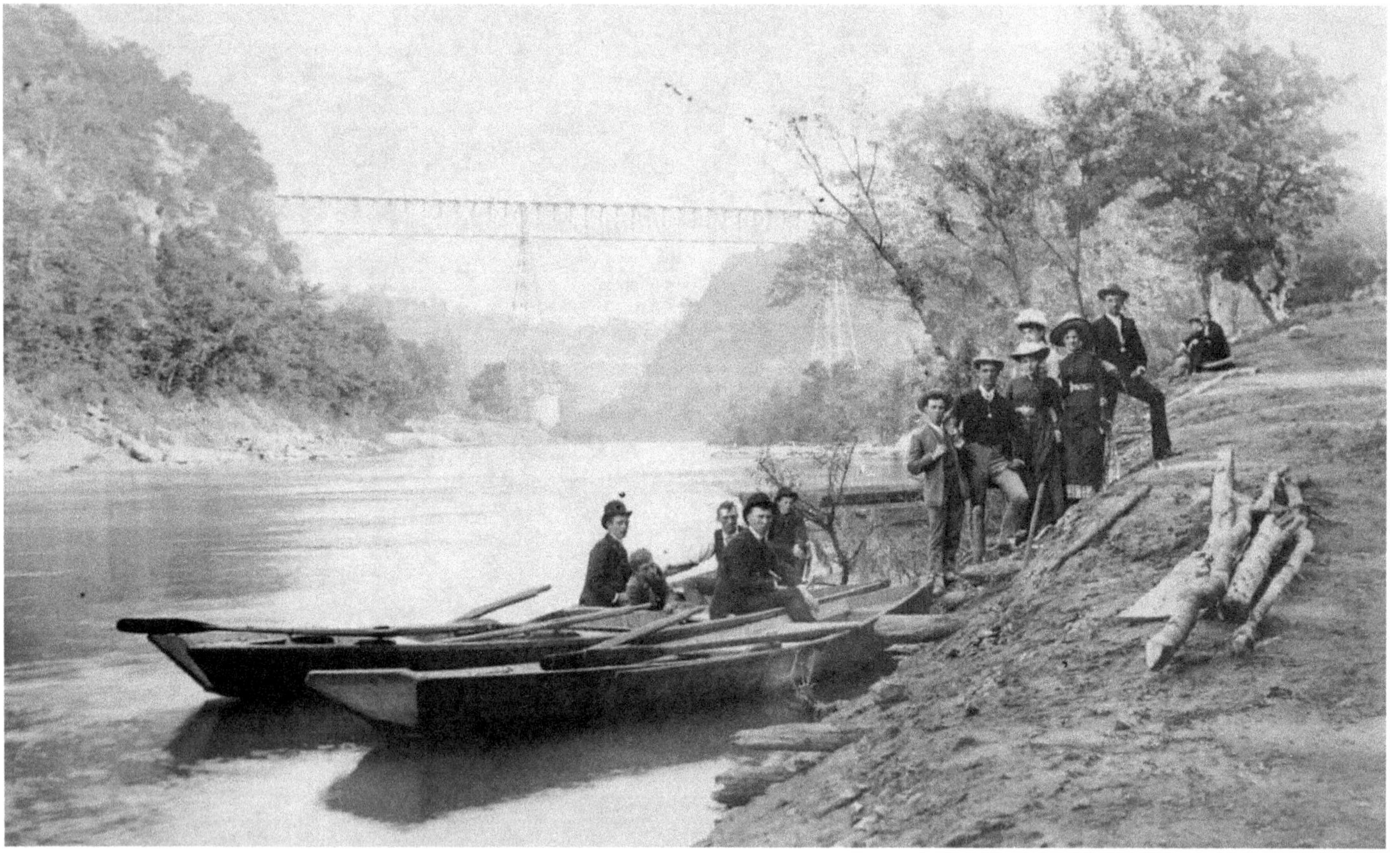

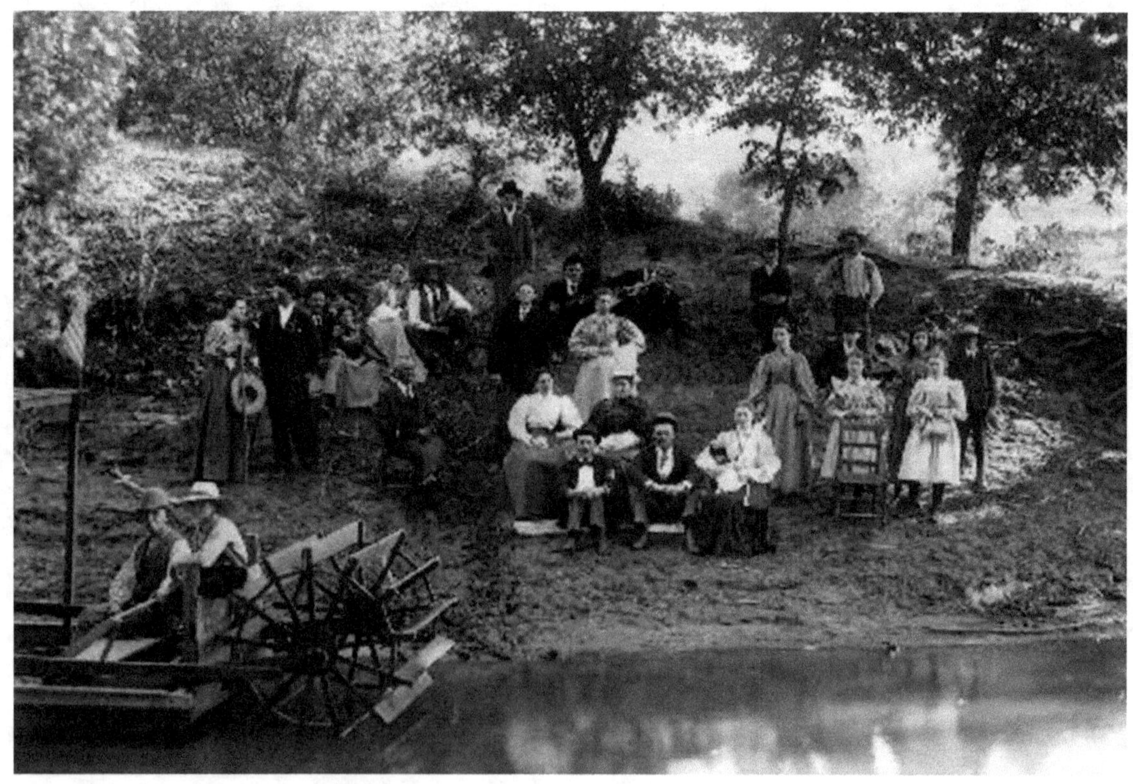

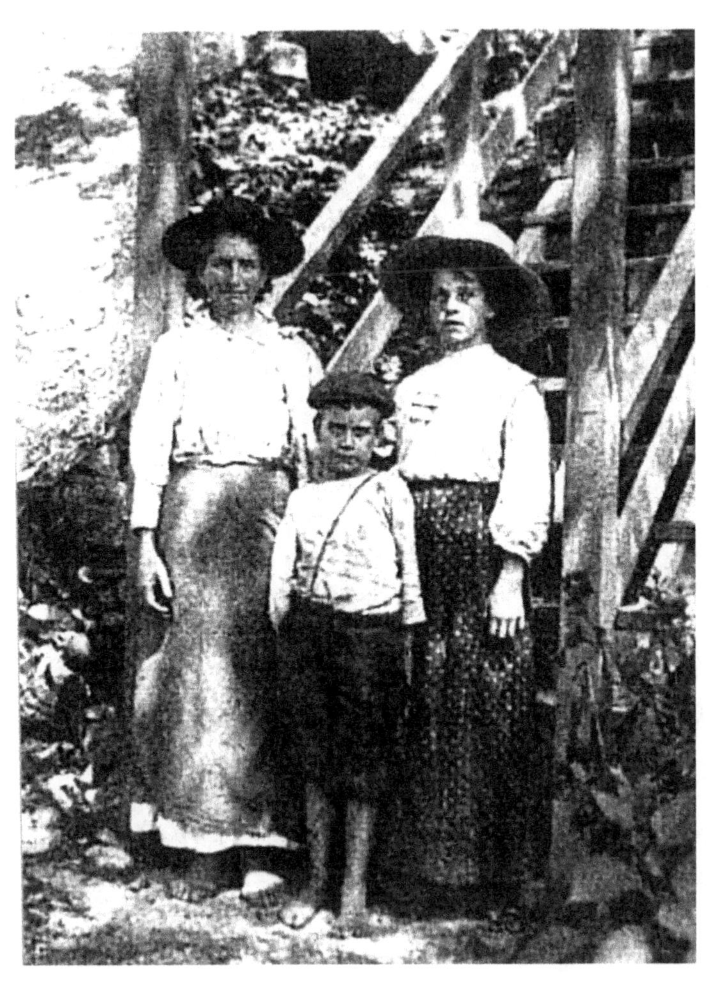

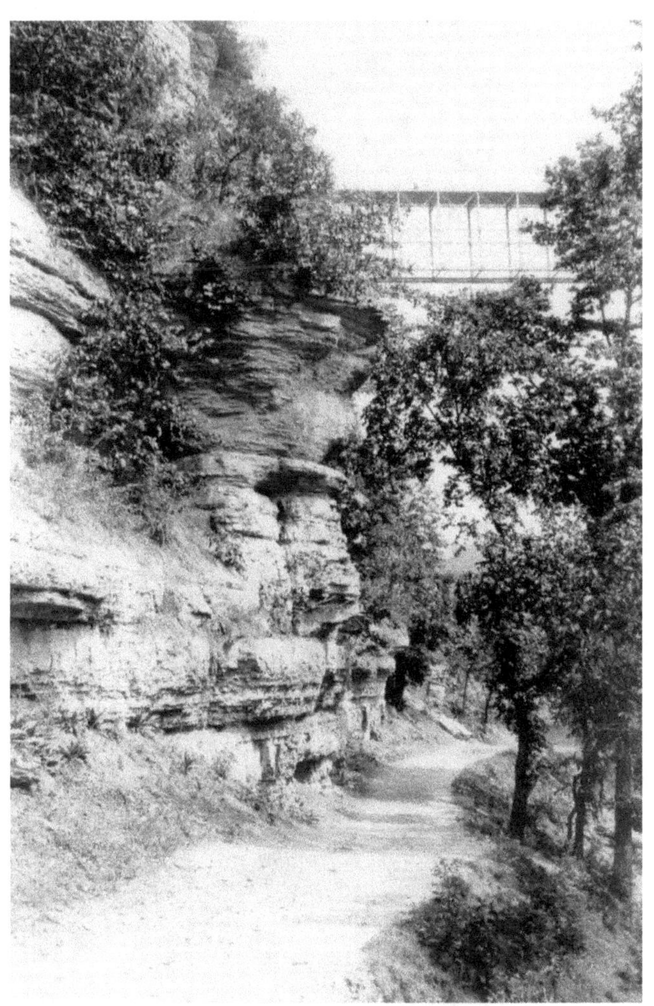

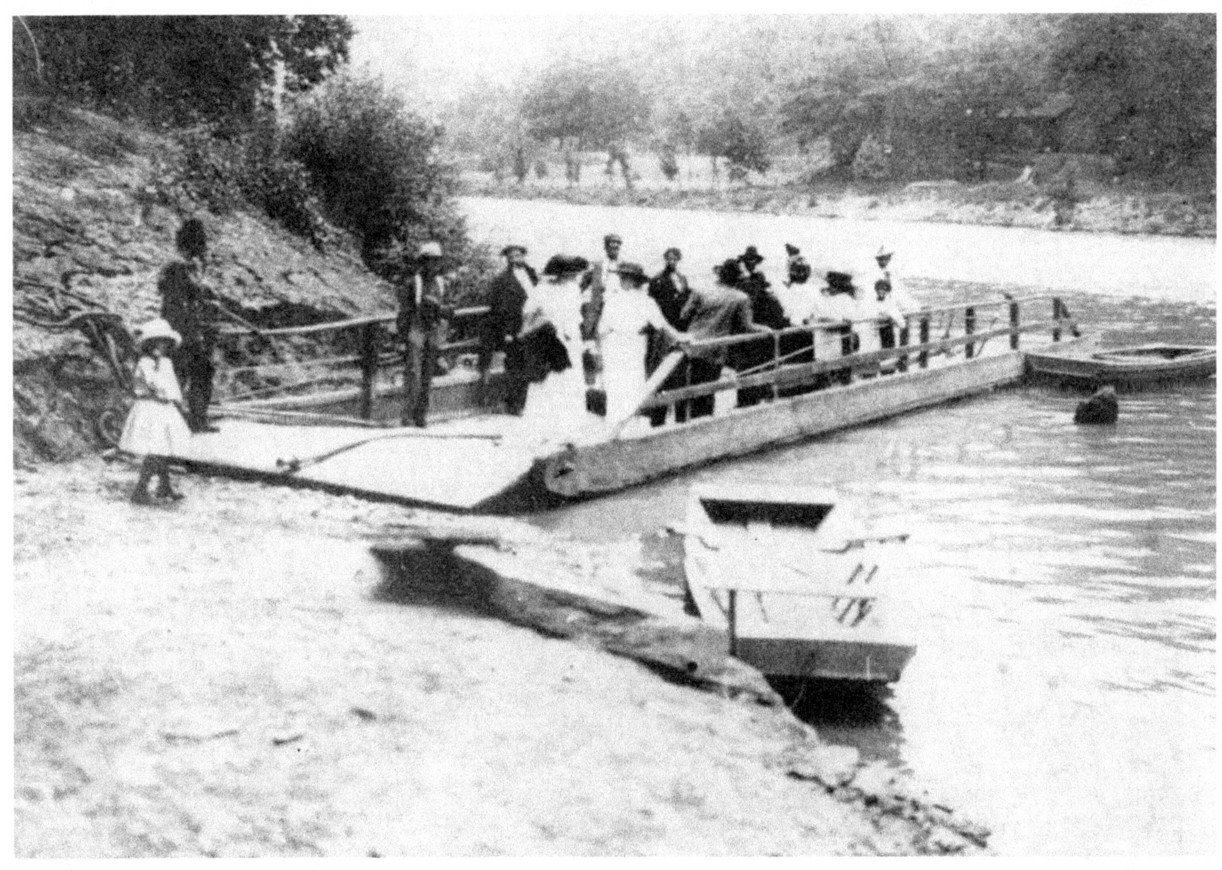

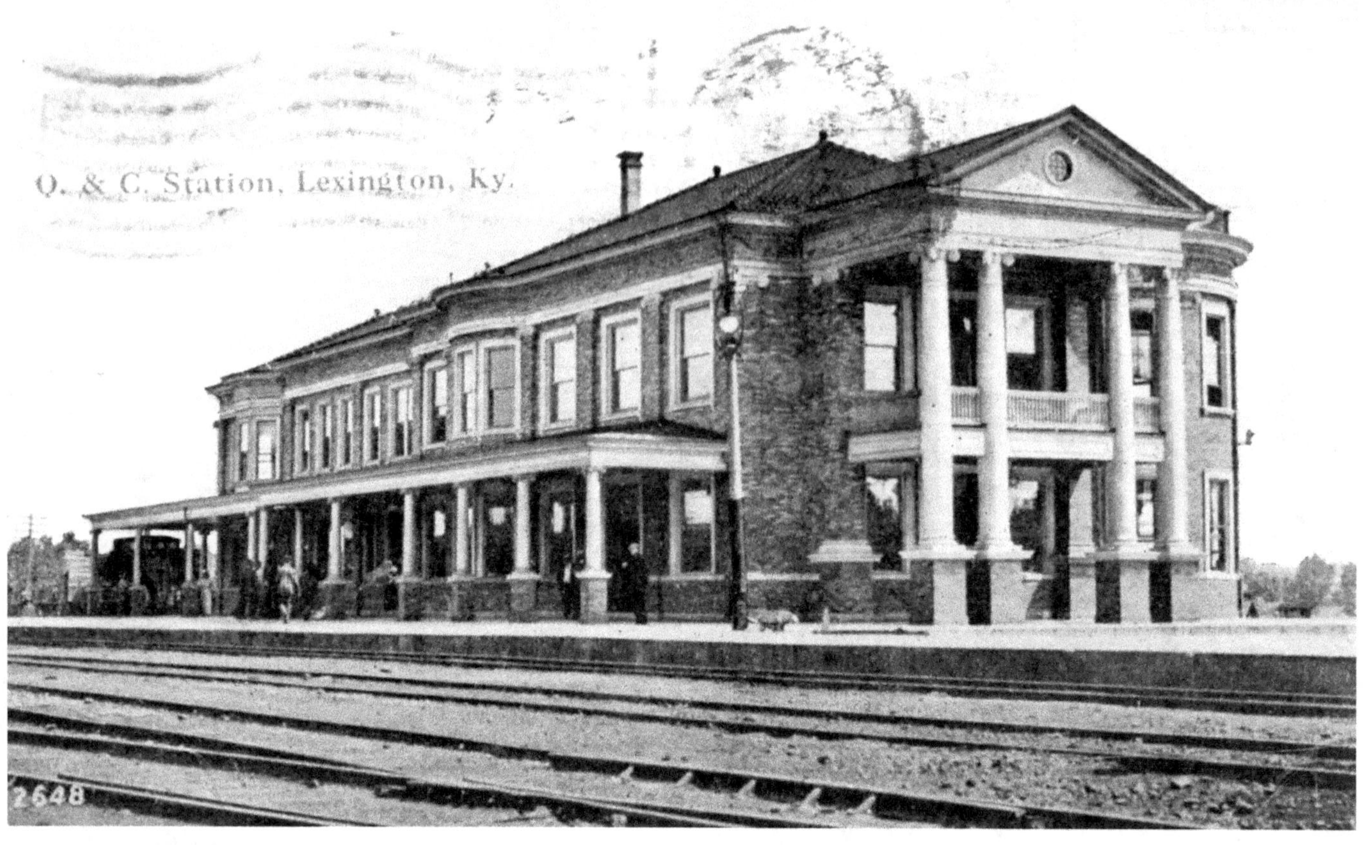

From and Queen and Crescent ad:

High Bridge on the Queen and Crescent Route is the most famous place known anywhere for a short excursion. A low rate of $1.00 from Lexington is made every week during the summer.

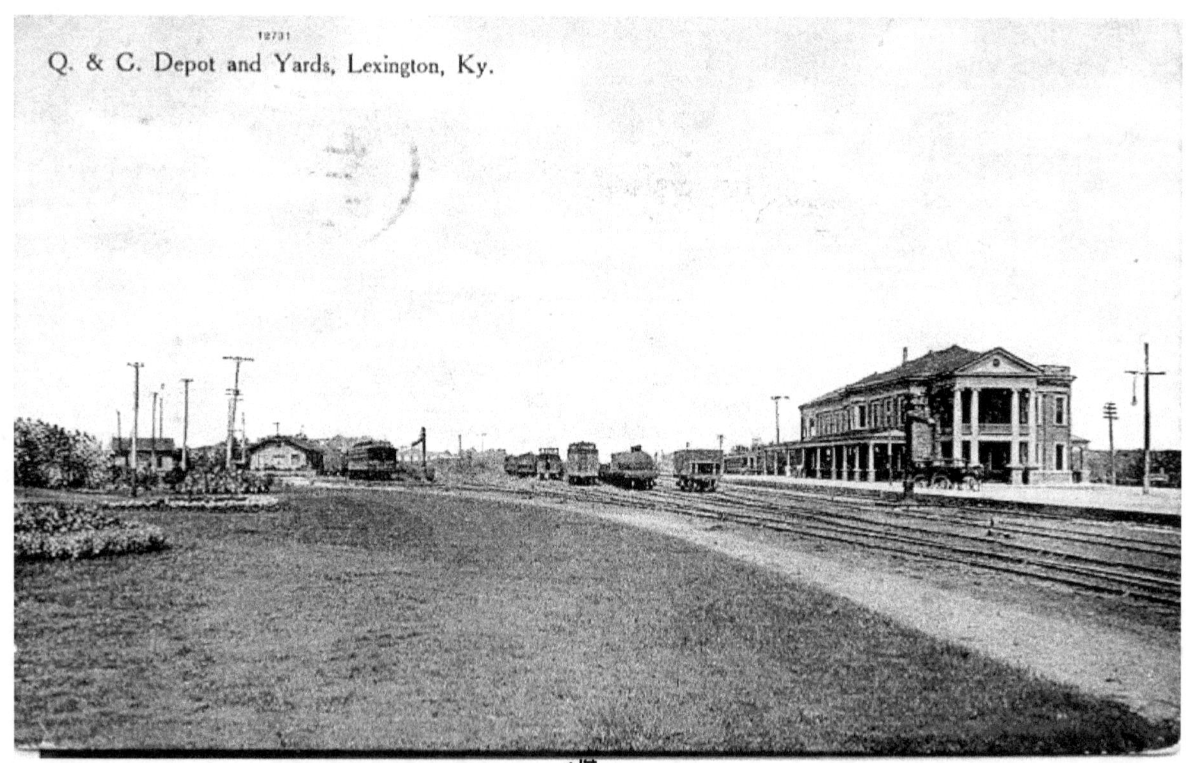

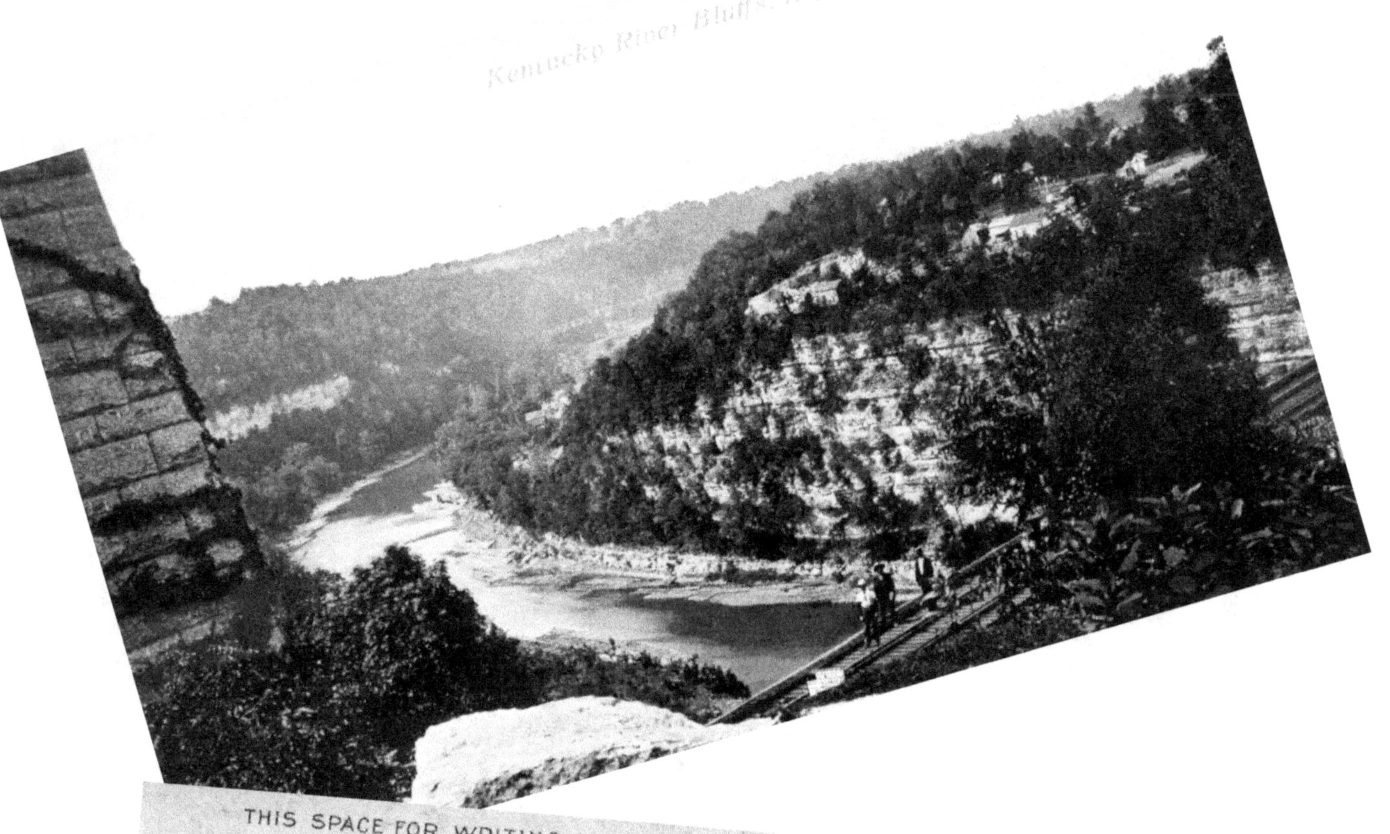

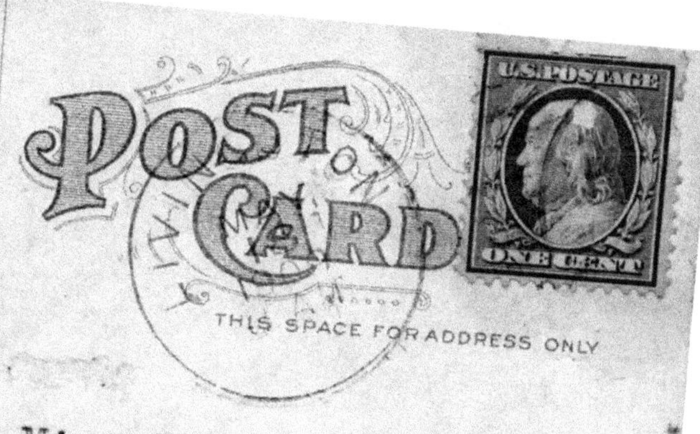
THIS SPACE FOR WRITING MESSAGES

WRENN & KING, IMPORTERS & MANUFACTURERS, LEXINGTON, KY. PHOTO BY KNIGHT.

Home again.Found your
good letter and book,can
never thank you enough,
have read the letter over
and over and over again.
Please write me early and
tell me how many and what
cards you have received
from me since I reached
Ky.and tell me how you
like them.Will write let-
ter soon.Tired and sleepy
now but well,had great
trip,wonderful country.
R.

POST CARD

THIS SPACE FOR ADDRESS ONLY

Miss Willie Fagan

Havana

Alabama.

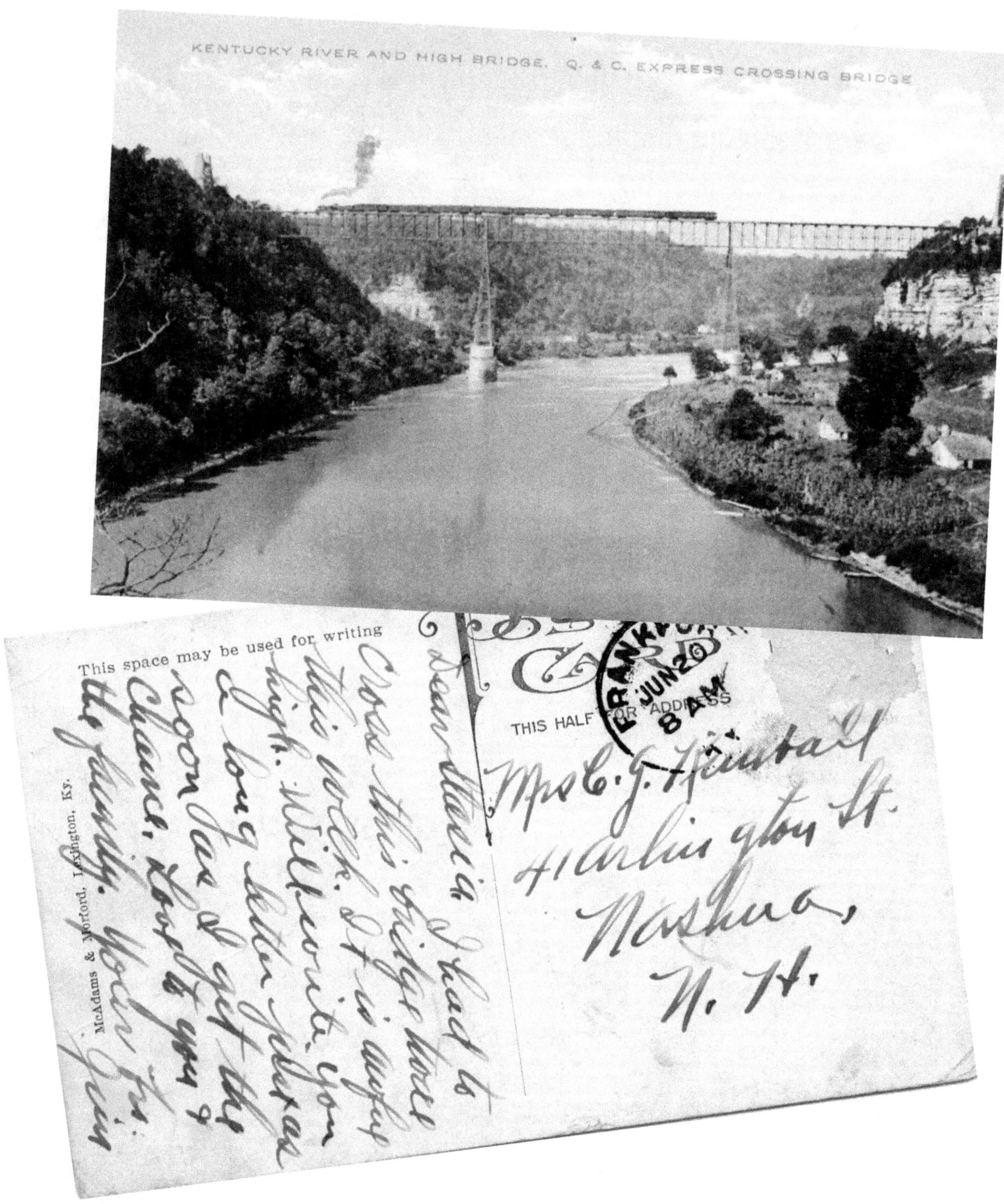

A popular means of communication around the turn of the century, many postcards were available depicting High Bridge on the front. Oftentimes after crossing the impressive structure, correspondence would mention the bridge. This 1908 postcard from "Jim" states in his opening line, "crossed this bridge twice this week. It is awful high," is typical.

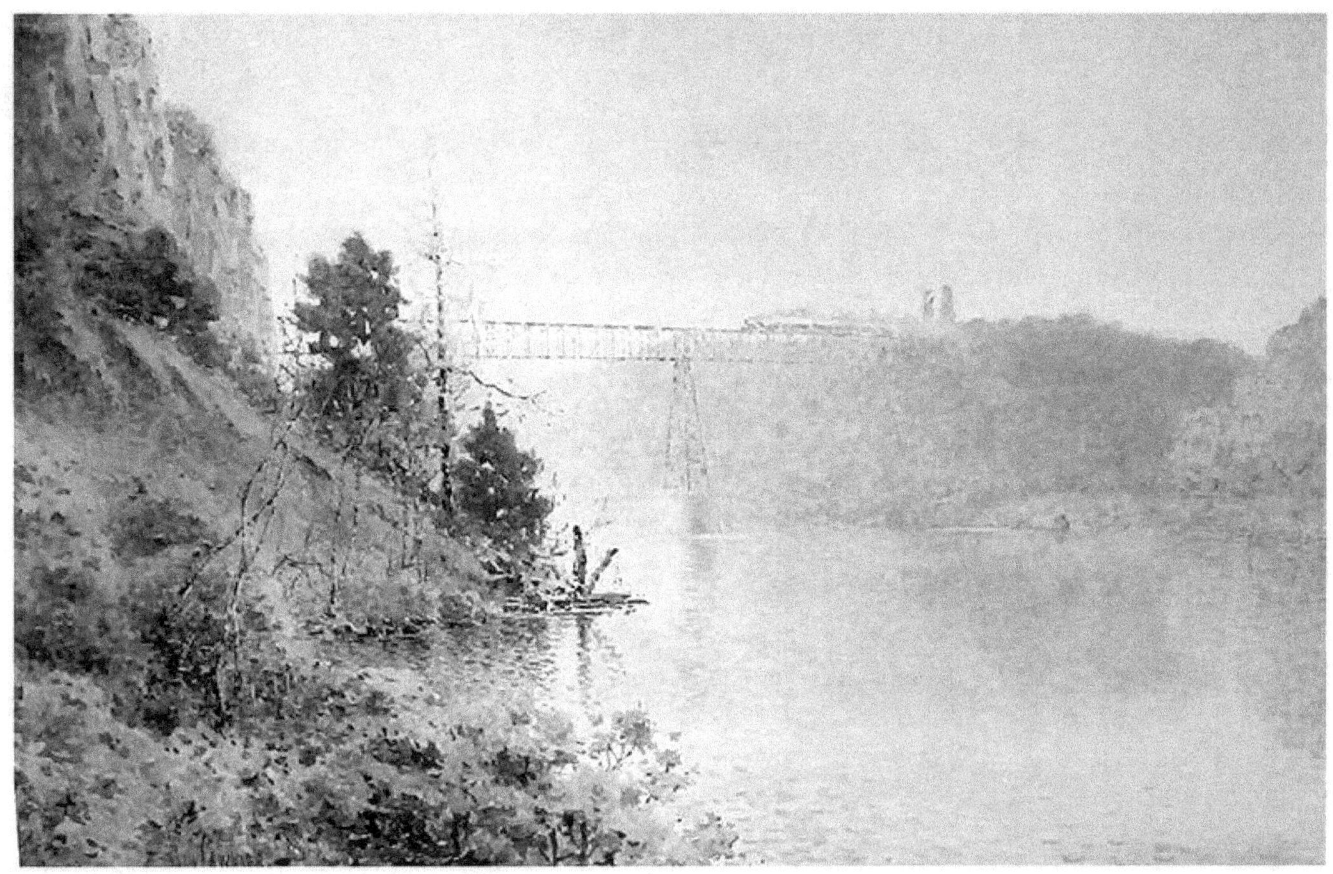

The beauty of the cliffs and the charm of the river captured the imagination of Kentucky artist Paul Sawyier, who lived in a houseboat on the river and painted scenes between Camp Nelson and High Bridge from 1908 to 1913. Even after relocating to New York in 1913, Sawyier continued to paint scenes from his beloved Kentucky River up until his death in 1917.

High Bridge often appeared in Sawyier's work. The above scene, simply entitled *High Bridge*, was created for his friend, Curtis Dougherty, the chief engineer of the Southern Railroad.

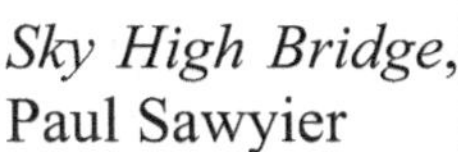

Sky High Bridge,
Paul Sawyier

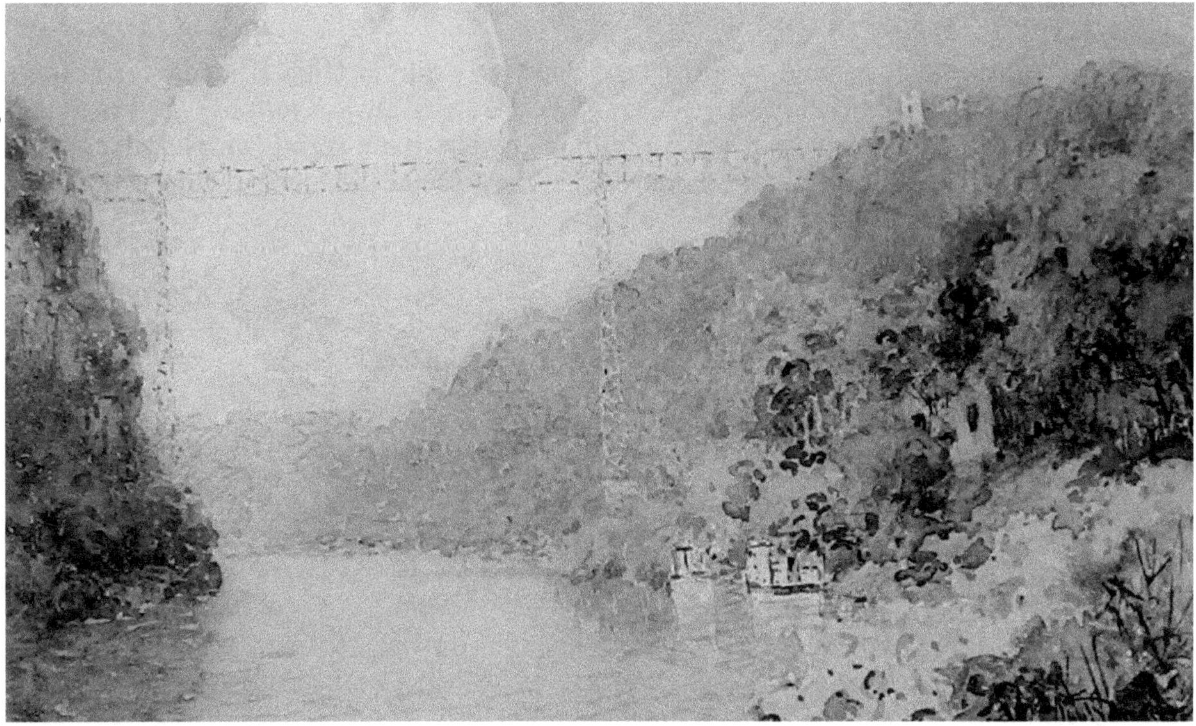

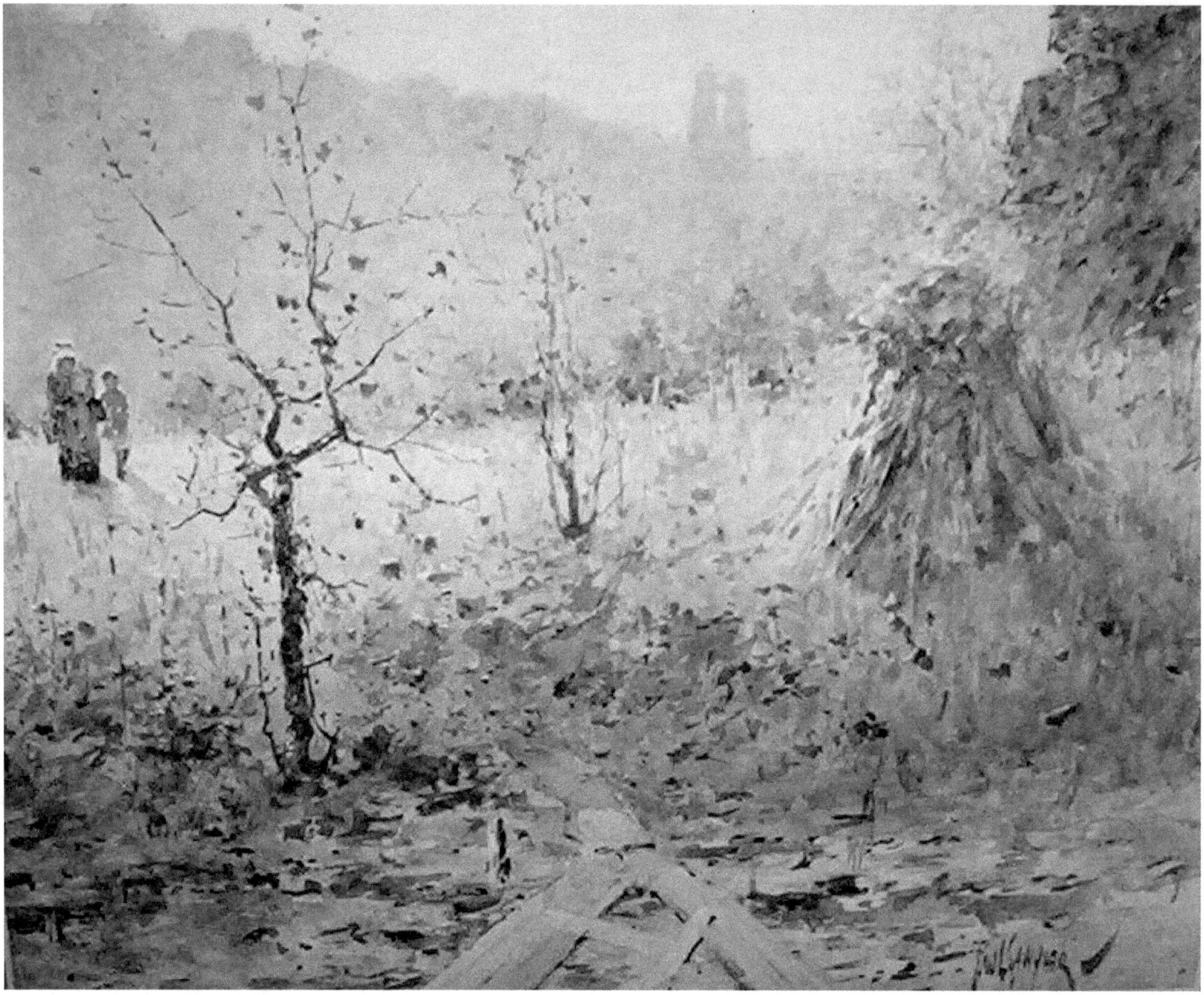

Going to the Spring was painted by artist Paul Sawyier while in High Bridge. According to Ken Houp, his great-grandmother, Joanna Dearinger Beckham, said Sawyier always told her he was going to include her in one of his paintings, but she did not believe him. On a visit to the spring with her daughter, Hadgie Beckham Horn, Sawyier was there painting. It is believed that they are the woman and child in this painting.

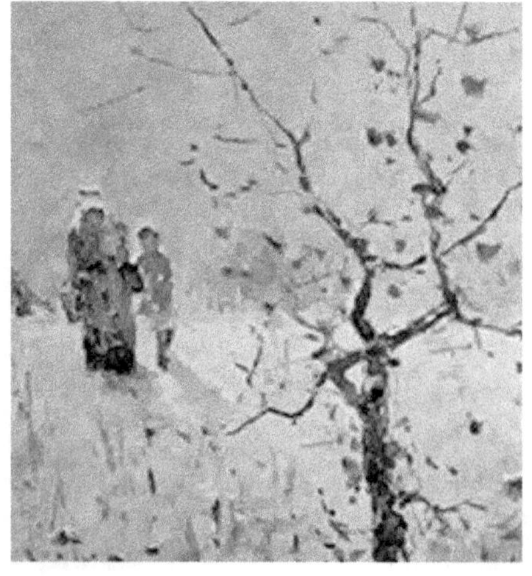

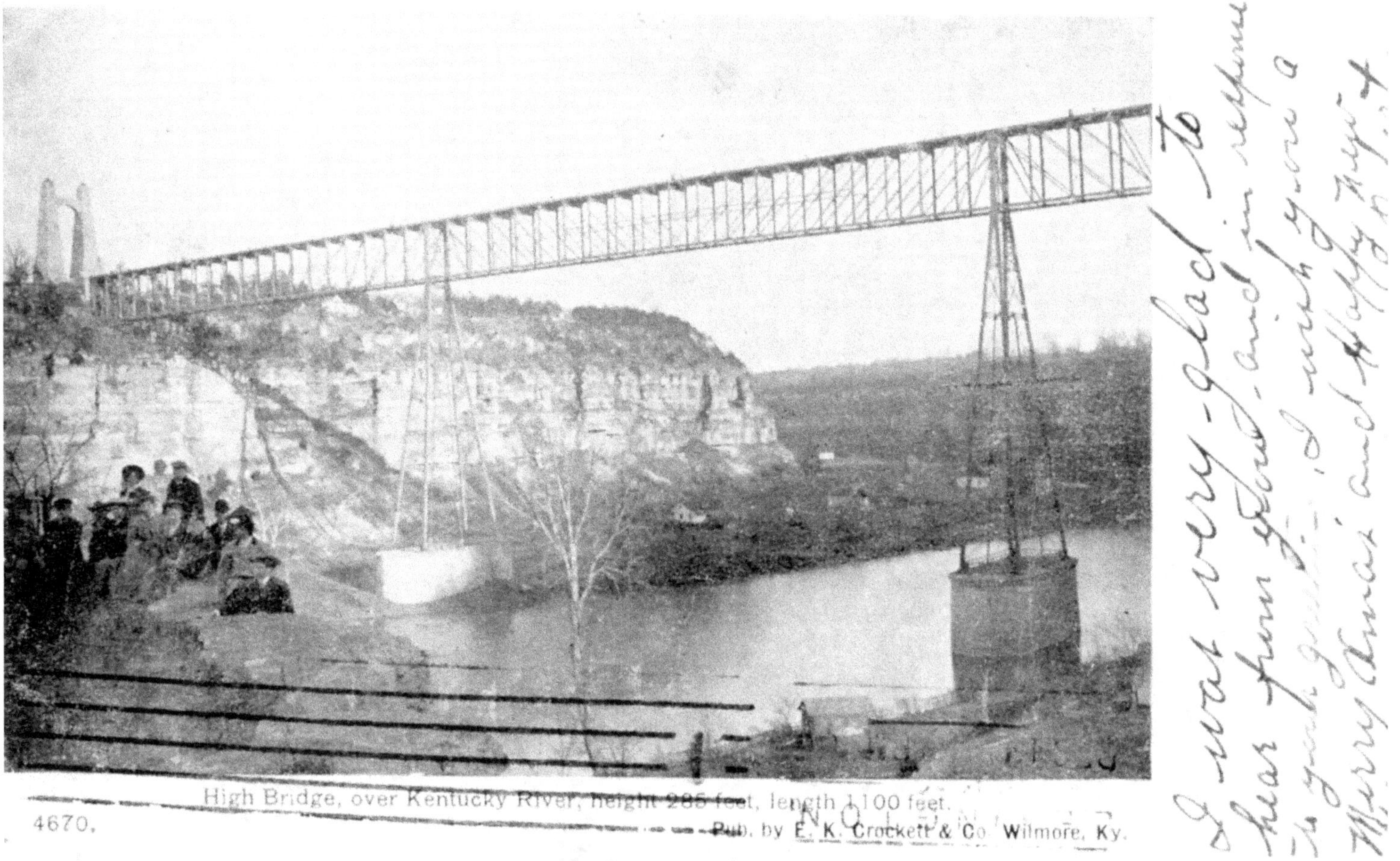

High Bridge, over Kentucky River, height 285 feet, length 1100 feet.

4670. Pub. by E. K. Crockett & Co. Wilmore, Ky.

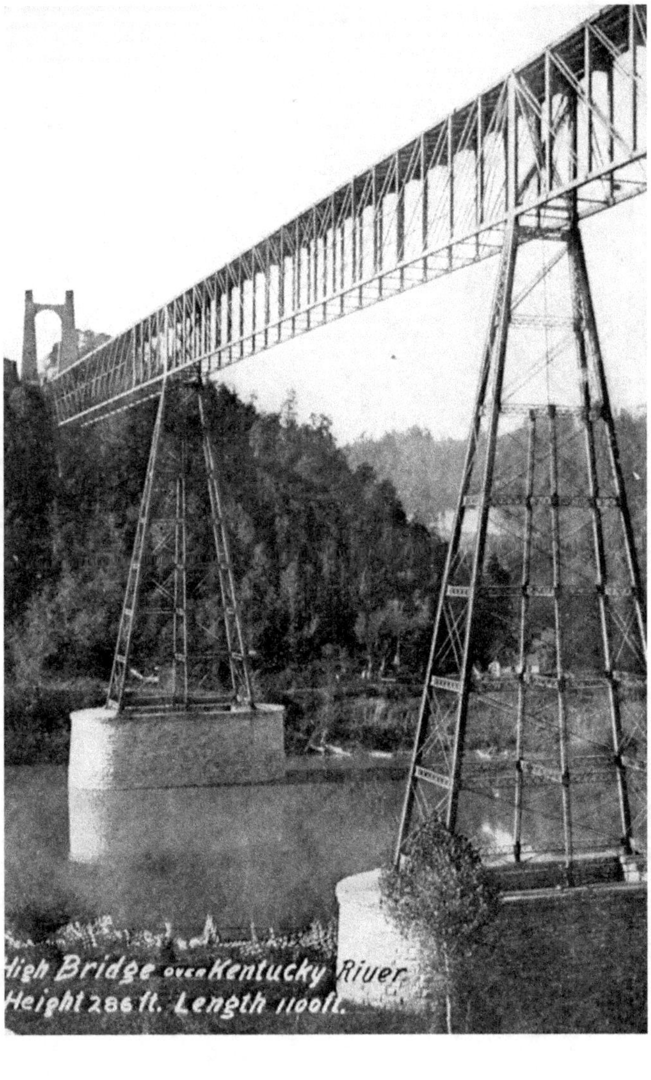

High Bridge over Kentucky River
Height 286 ft. Length 1100 ft.

Built just strong enough to accommodate the lighter rail traffic of its day, the original design of High Bridge was becoming outdated. Questions arose about the safety of the structure. These concerns signified the beginning of another era for the community that would change its face forever.

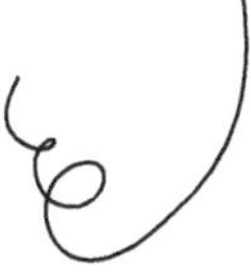

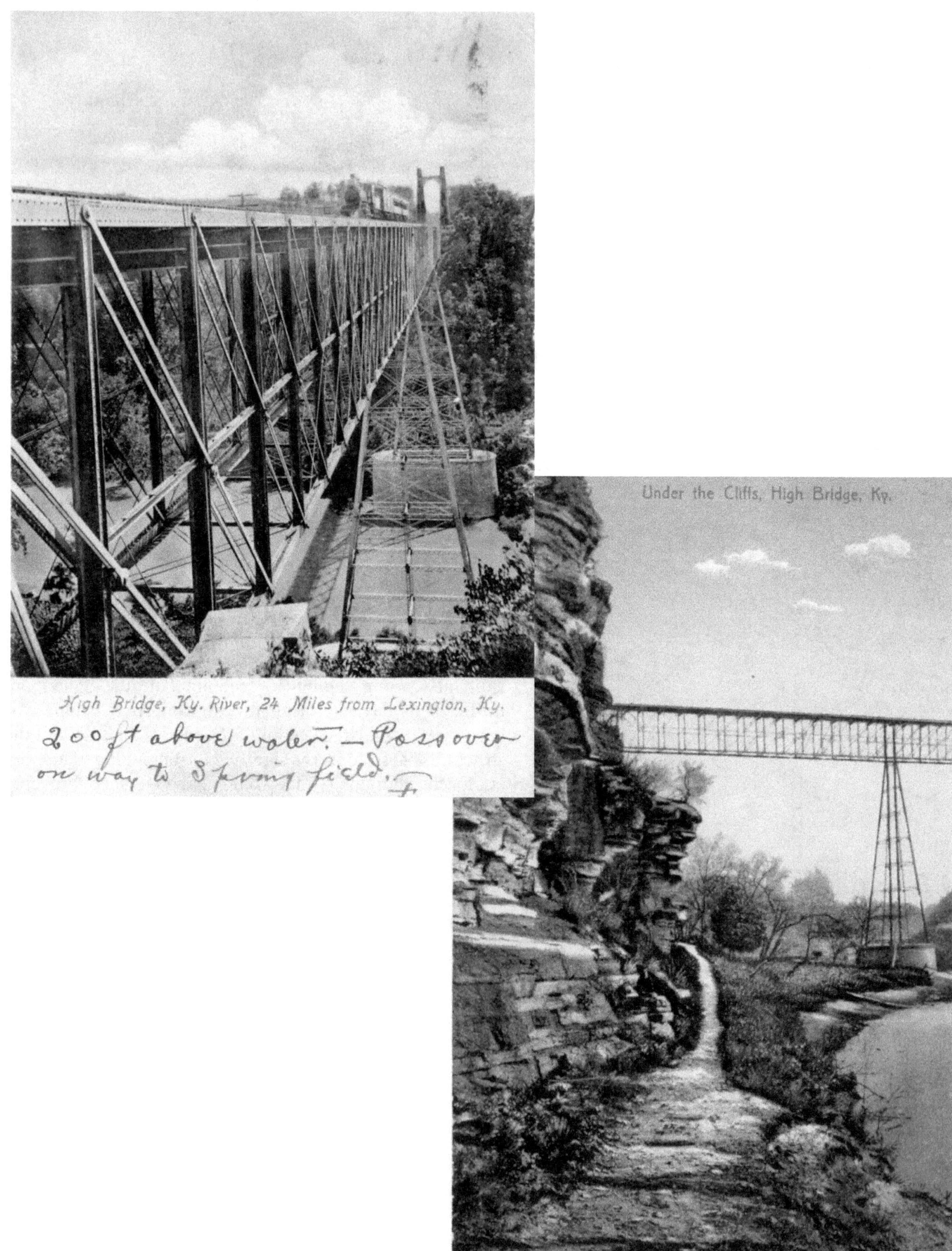

High Bridge, Ky. River, 24 Miles from Lexington, Ky.

200 ft above water. — Pass over on way to Springfield. F

The New High Bridge

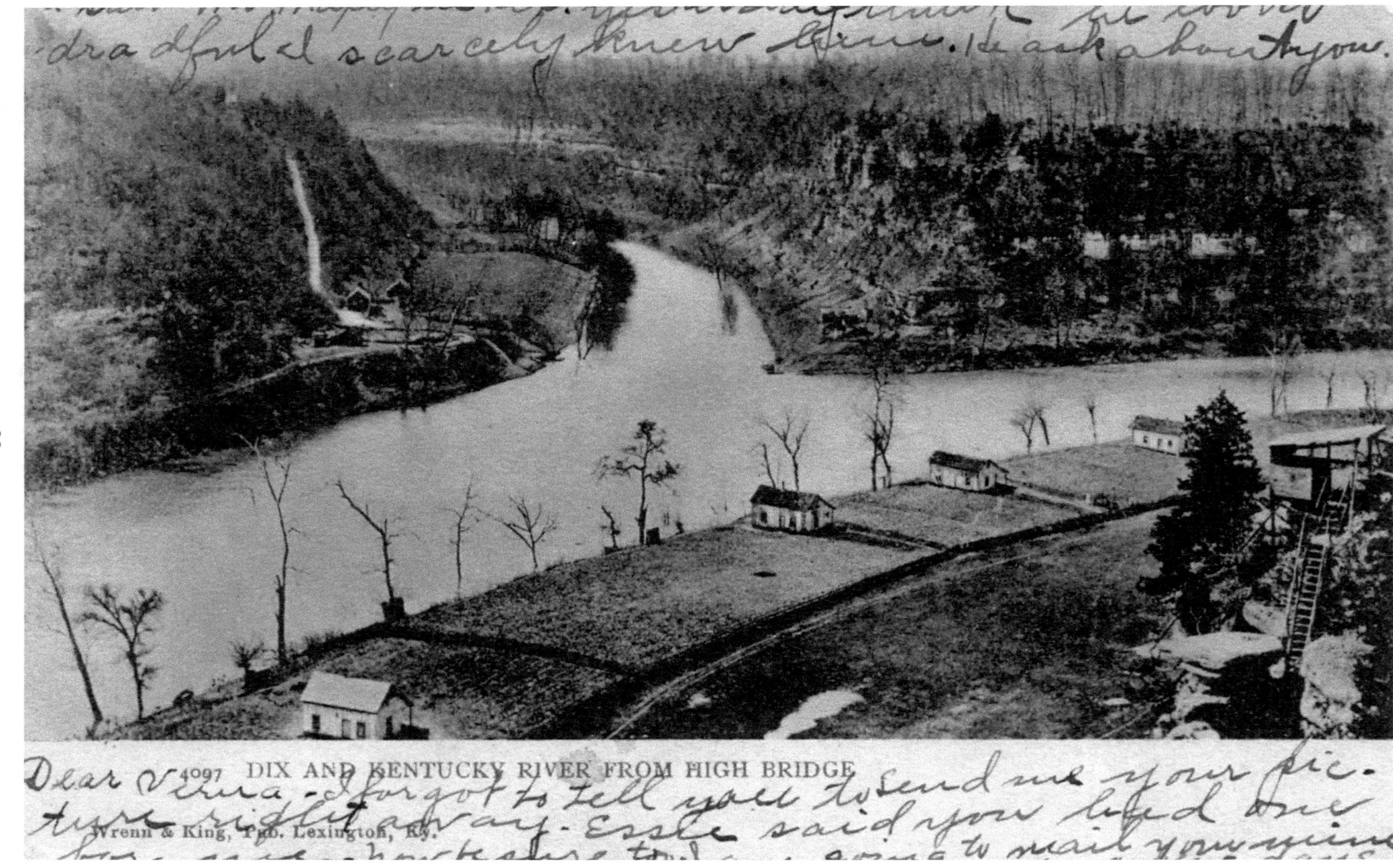

The producer of this postcard mistakenly printed the image backwards!

When Charles Shaler Smith designed High Bridge in 1876–1877, he could not have envisioned the heavy trains with increasingly heavy loads that would cross the bridge on a day to day basis. It was apparent by the early 1900's that the old bridge was not up to the job, and a project was undertaken to build another bridge. The American Bridge Company of New York was contracted for the job under the supervision of Curtis L. Dougherty, the chief engineer of the Southern Railroad. However, like the original builders of the bridge who were faced with the challenge of the cliffs, these engineers faced their own challenge: how to replace the bridge without interrupting rail travel. An interesting solution to the problem was devised by engineer Gustav Lindenthal. They built the new structure *around* the old bridge. Work began on the area around the bridge in 1908, and work on the bridge itself started in 1910. The new High Bridge was completed in 1911, and the first train crossed the bridge amid great fanfare.

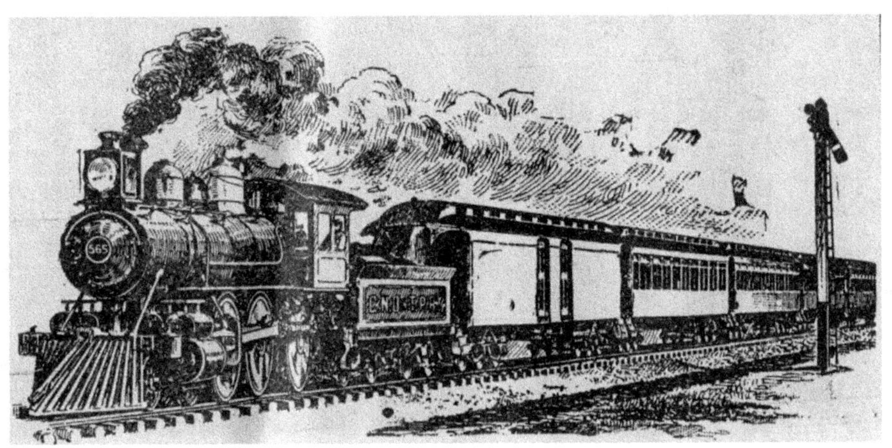

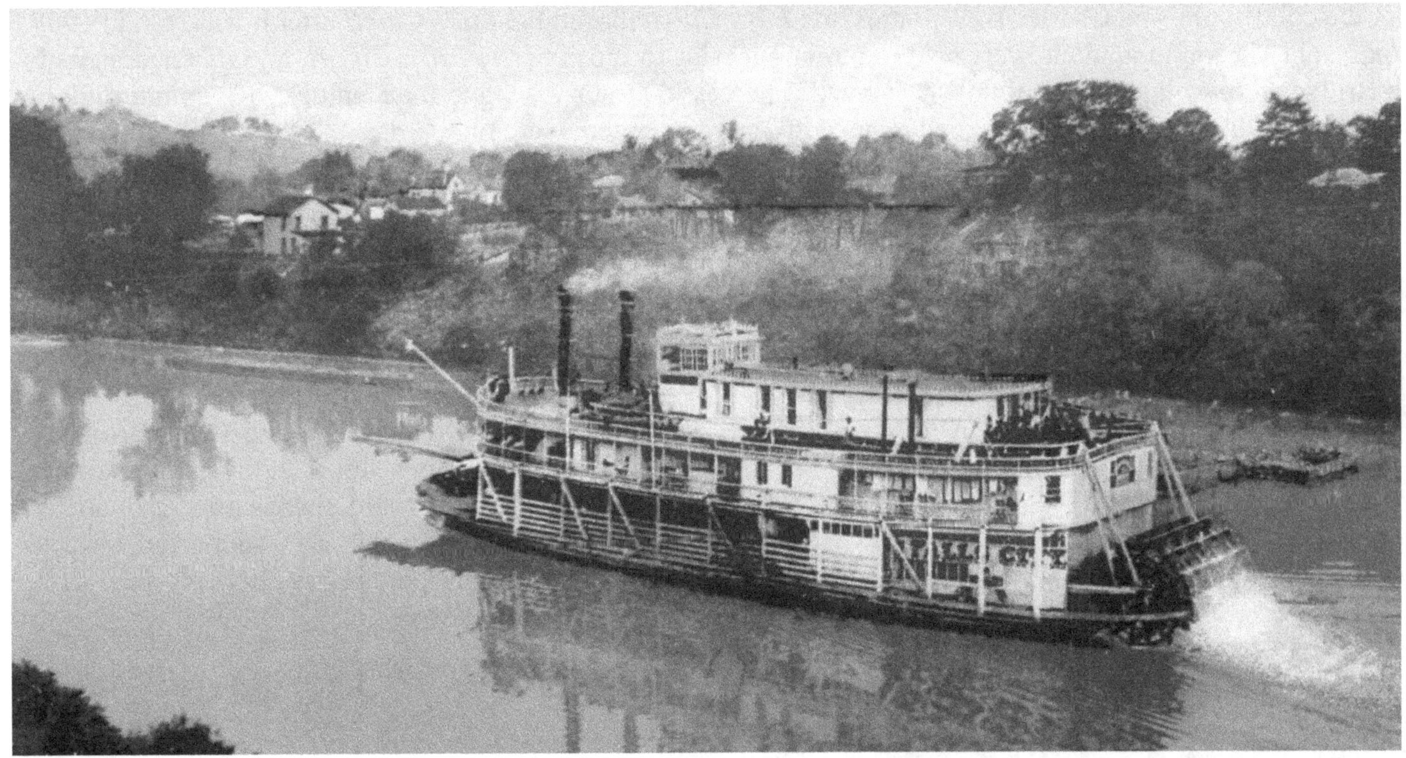

Much to the regret of many, due to a decline in demand for riverboat cargo transport and the fact that the profitable excursions on the river could not be held year-round because of weather conditions, the *Falls City II* was sold in February 1908 to Capt. Thomas Morrissey. He took her to Vicksburg, Mississippi, to replace his vessel the *Rosalie M*, which foundered in a storm in Mississippi some weeks before. The last sighting of this majestic boat that was once the best on the Kentucky River was in 1916, when it was seen abandoned in a Mississippi shipyard, falling apart.

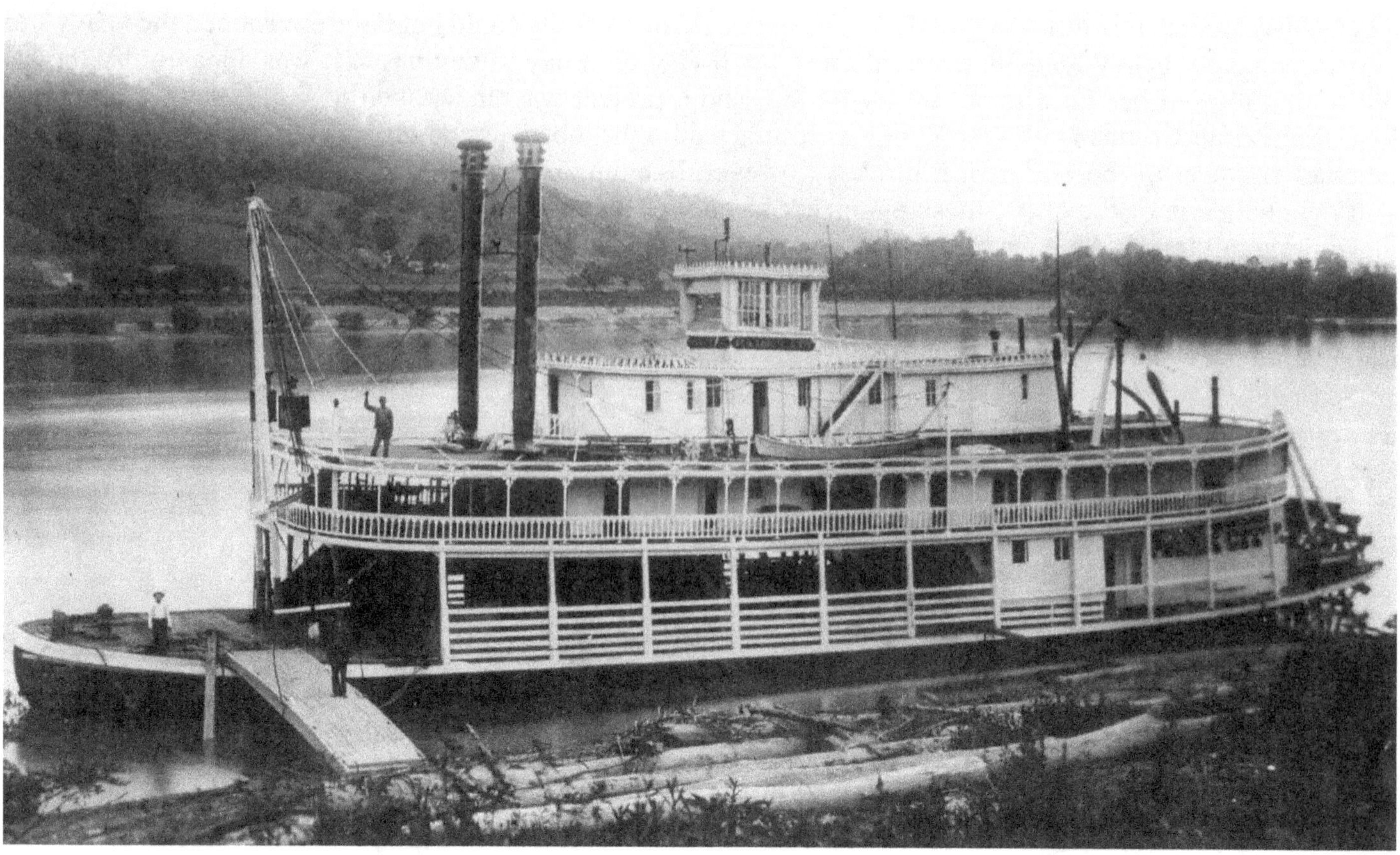

Under the control of Capt. Charles B. Williams, the *Park City* replaced the *Falls City II* after it was sold in 1908. The passenger accommodations were said to be far superior to those of the *Falls City II*, and its excursions to High Bridge were well attended, if not short-lived. The vessel struck a snag in the Kentucky River and sank on December 9, 1909. Fortunately, the crew escaped and no passengers were onboard.

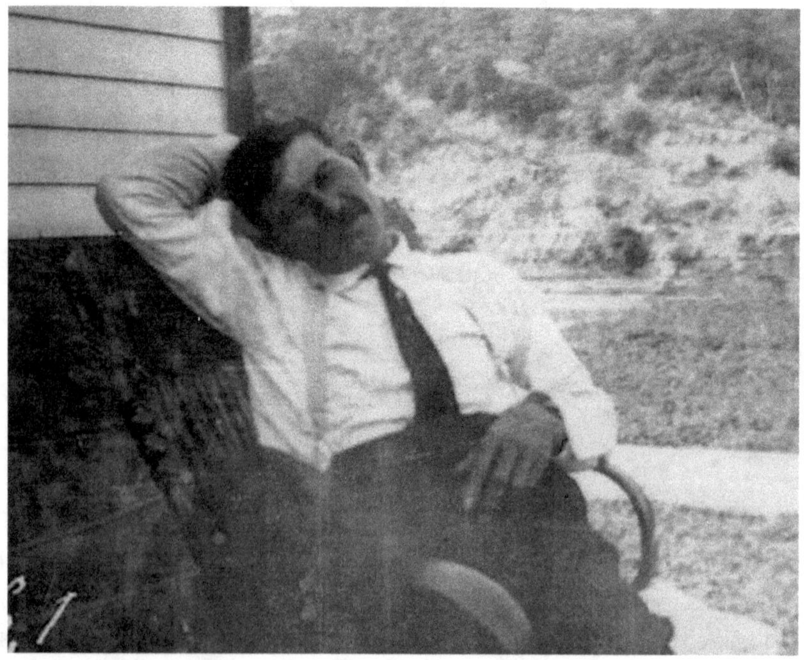

Seen here relaxing at home in High Bridge, James Robert Preston, brother of Capt. Squire Jordan Preston, operated a general store that served the High Bridge area for many years. J. R. Preston's motto was that he would always give each and all a fair deal. He also supported local farmers by purchasing their produce to sell in his store.

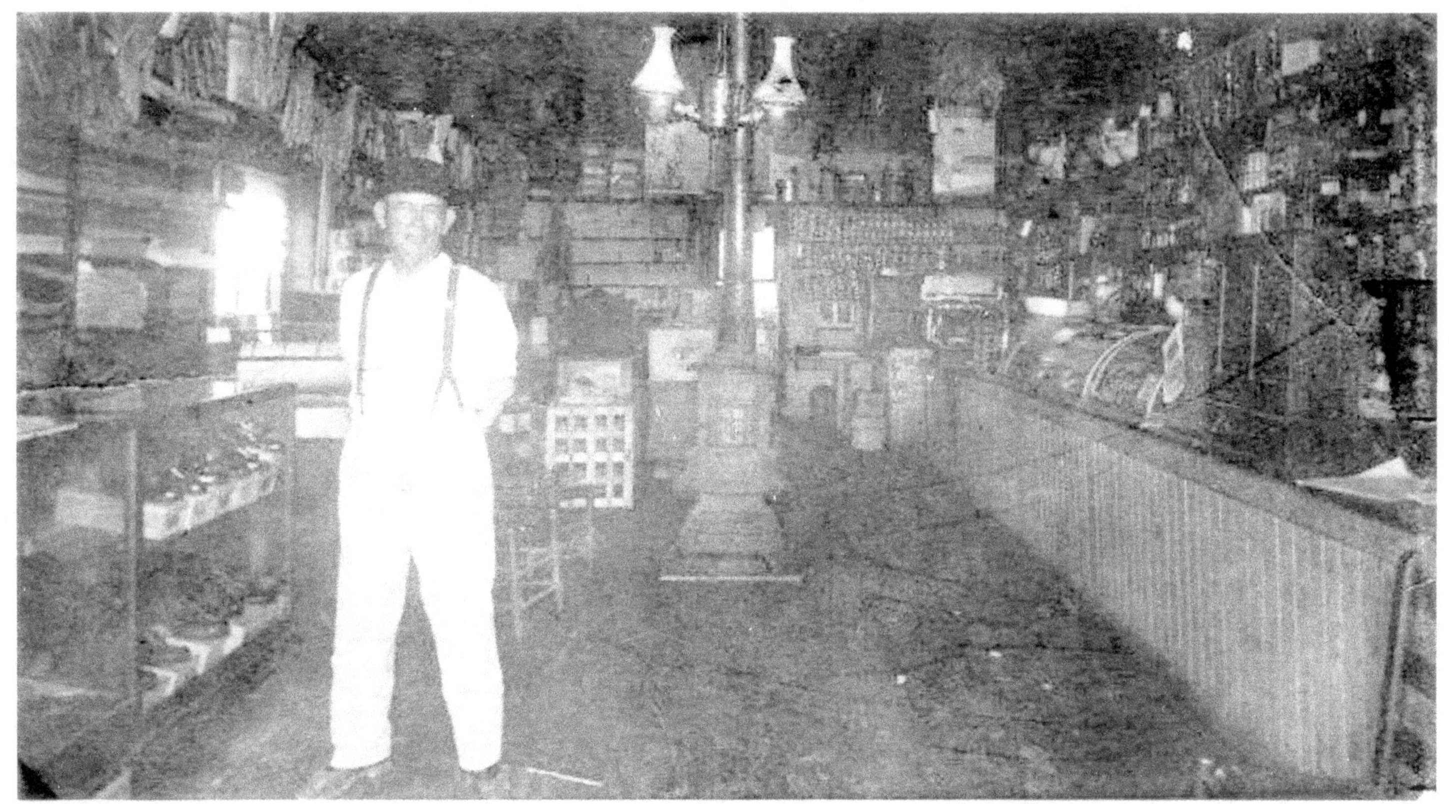

A May 26, 1909, *Jessamine Journal* newspaper mention of Preston's General Store stated, "J. R. Preston desires to announce to his many friends and customers that he has received, and is opening a large stock of foreign and domestic dry goods, notions, white goods, shoes, hats, etc. for the spring trade and he desires them to call and inspect same. He will quote prices most tempting however, this doesn't mean he has in any way abated his attention to his line of the best staple and fancy groceries." The man in the photograph is unidentified.

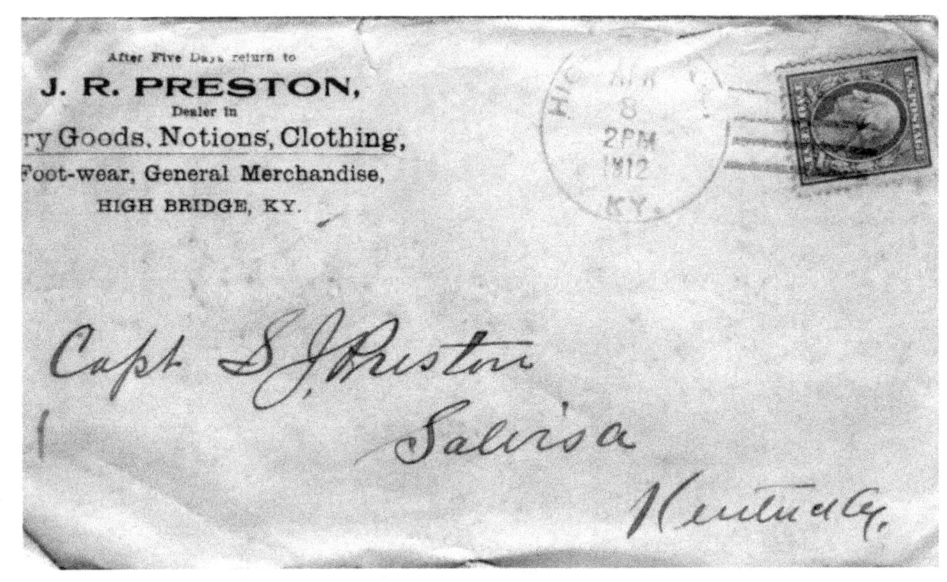

Advertising "Dry goods, notions, clothing, foot-wear and general merchandise," this envelope postmarked in High Bridge on April 8, 1912, from J. Robert Preston is addressed to his brother, Capt. Squire Jordan Preston, who lived in nearby Oregon. However, by 1912, the post office in Oregon closed and all mail was delivered to Salvisa, Kentucky, instead.

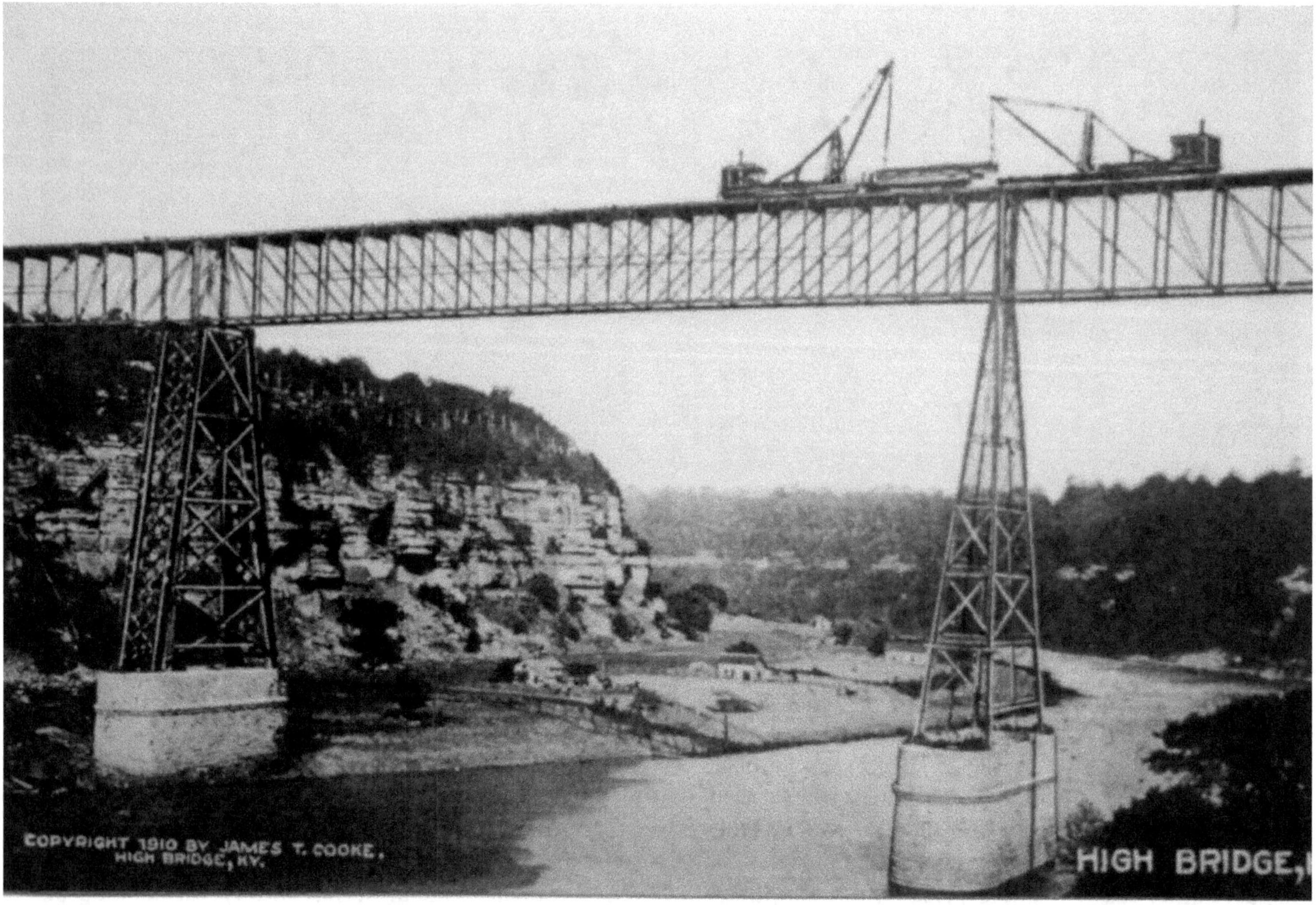

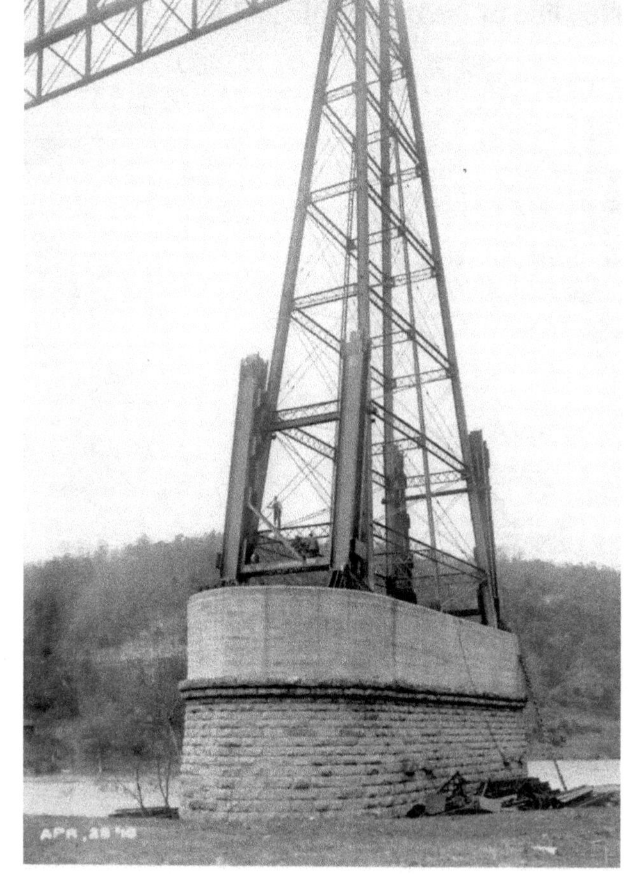

Before the new bridge could be built, its existing stone and steel piers had to be reinforced with additional steel and concrete to support the weight of the new bridge. Work began on the stone piers on April 24, 1910. In the above view from May, the built-up stone piers have been completed, and two cranes are atop the original bridge in preparation for the work on the reinforcement of the steel bridge support piers.

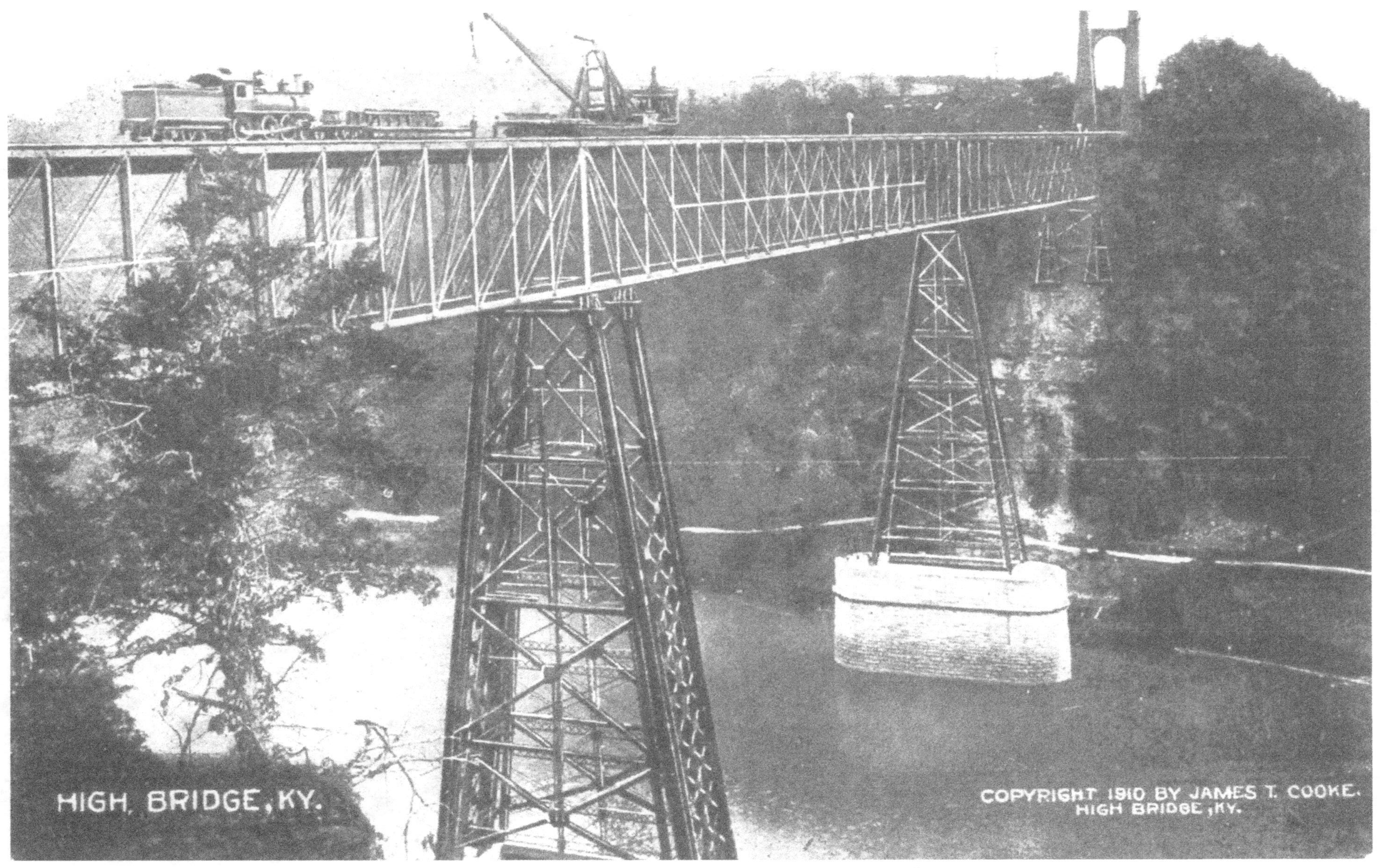
HIGH BRIDGE, KY.
COPYRIGHT 1910 BY JAMES T. COOKE.
HIGH BRIDGE, KY.

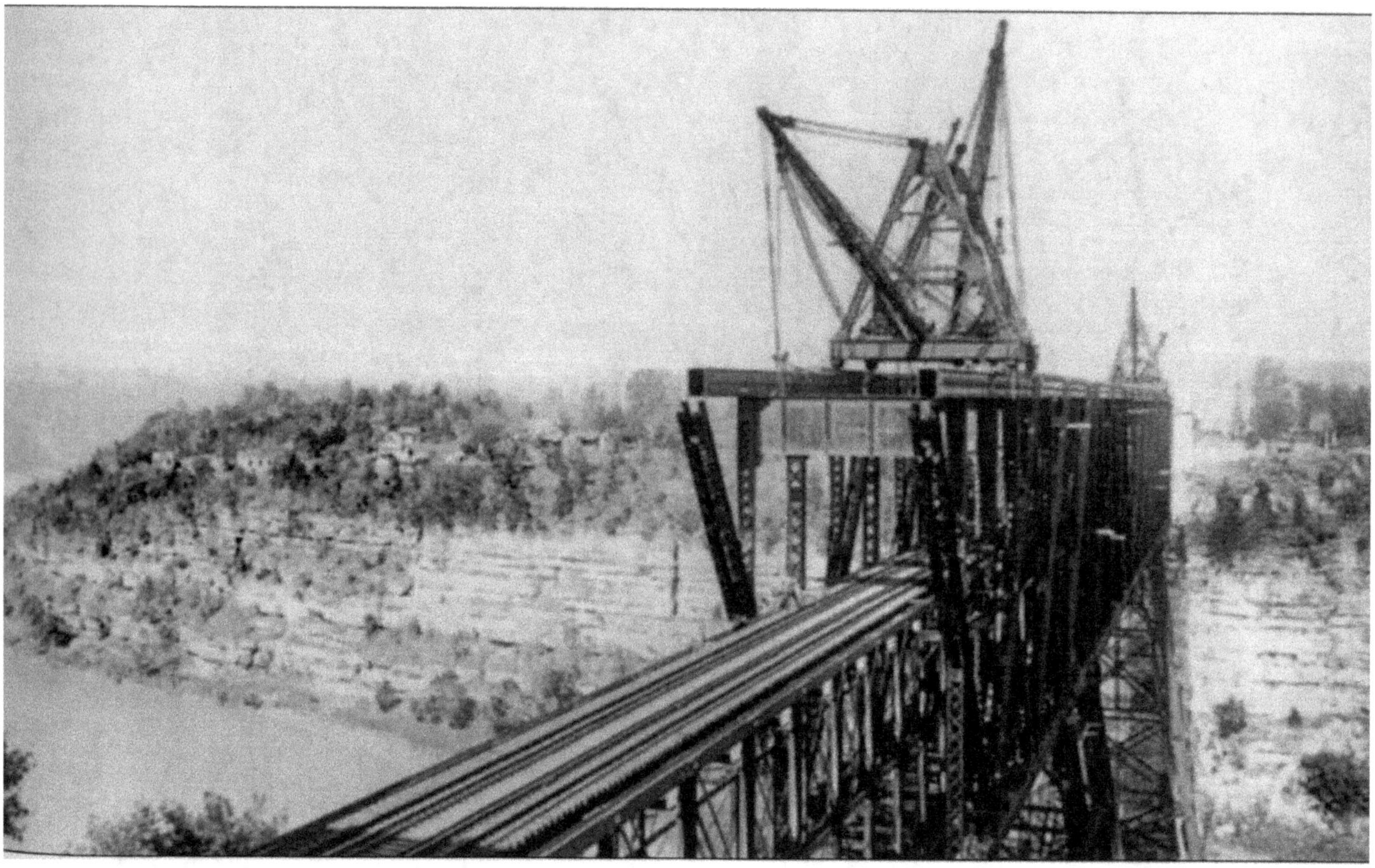

These views show the unique method of the construction of the new bridge around the old bridge. Until the opening of the new bridge in September 1911, trains would cross the old bridge while work on the new structure was going on above them by the American Bridge Company. The new bridge's track would be thirty-one feet higher than that of the old bridge and wide enough to accommodate a double track, should one become necessary.

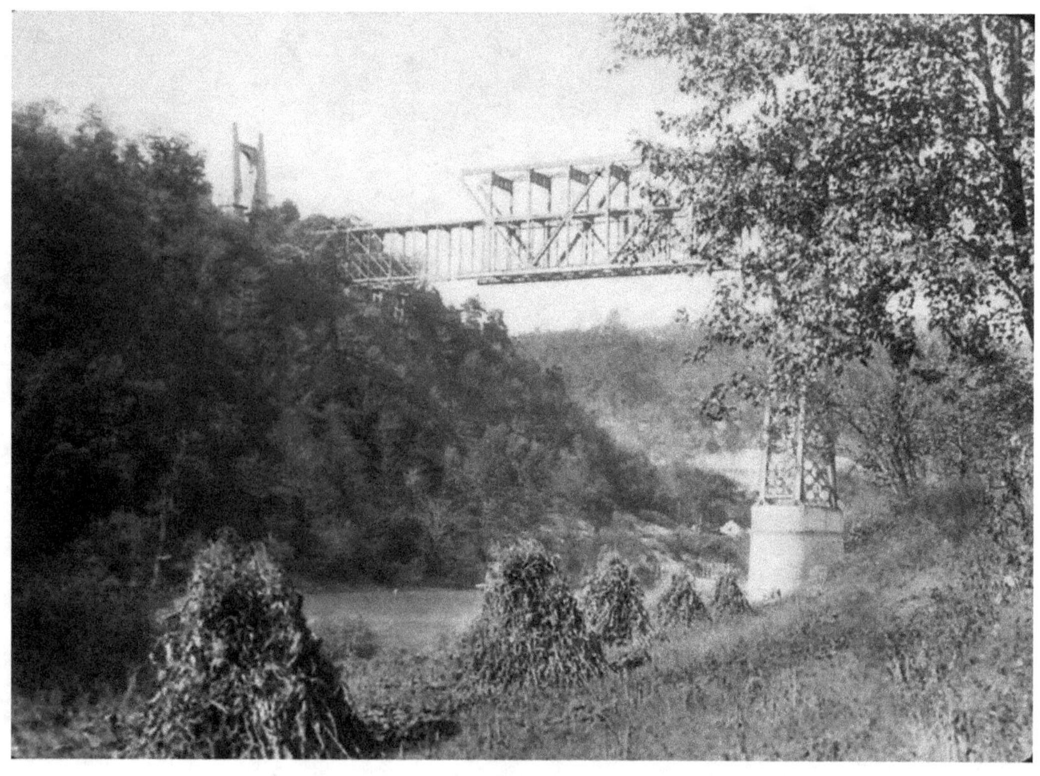

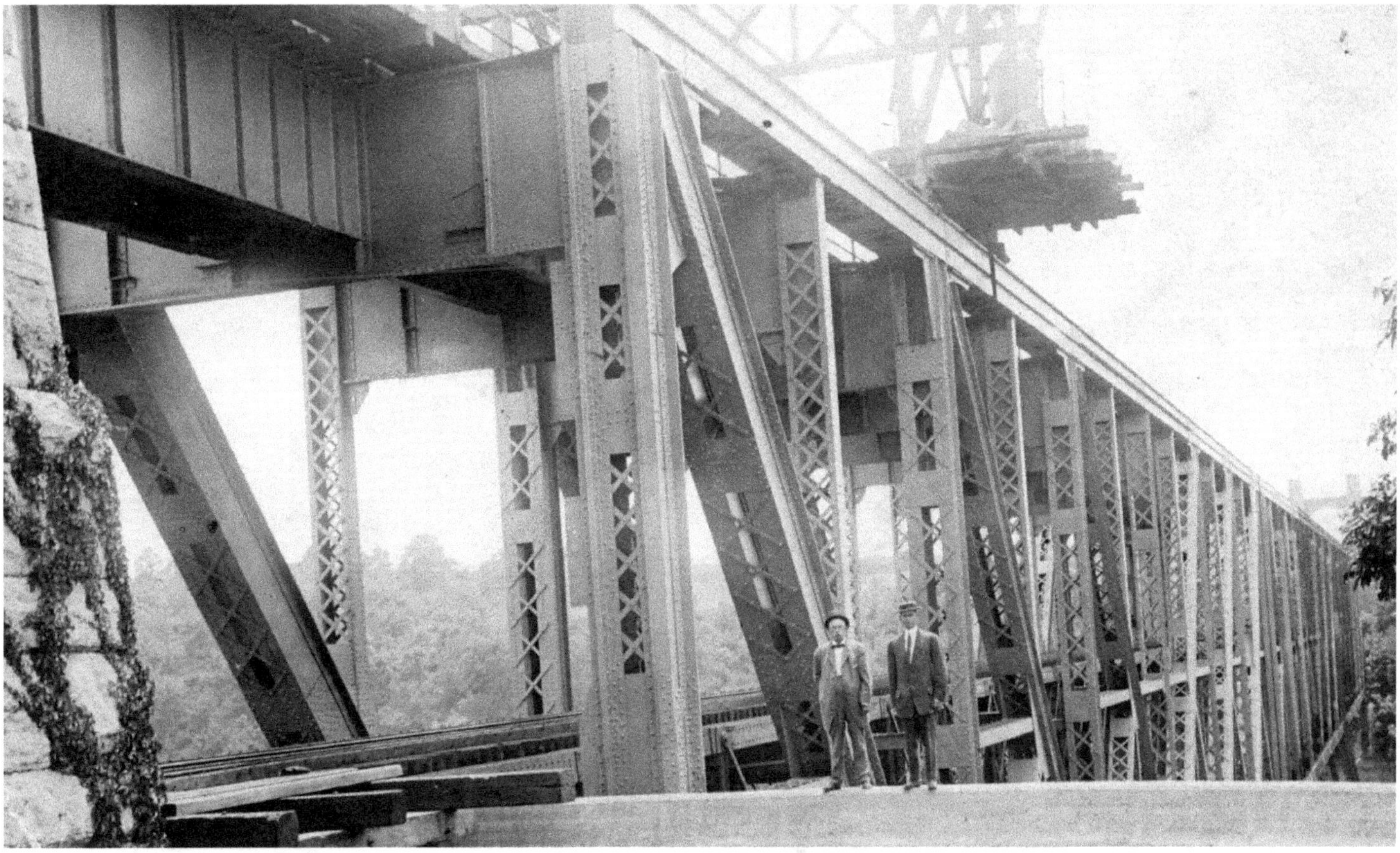

These two gentlemen are perhaps part of the group of 25 students who came from the College of Mechanical Engineering from Kentucky State University on October 28, 1910, to High Bridge. They observed the construction of the new bridge, standing atop the concrete wall that one sees today, which runs on the side of the roadway underneath the bridge at the top of the cliff along the cliff edge.

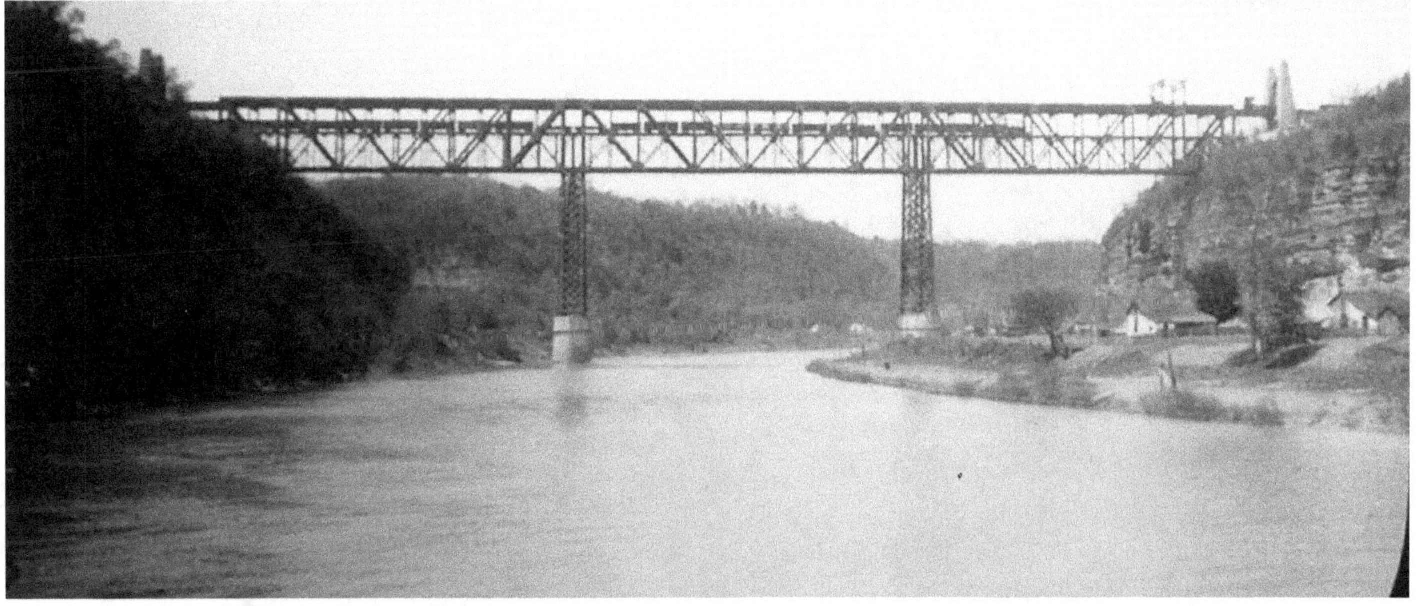

Nearing completion, one can appreciate in this distant view the train that goes across the old structure inside the new bridge. Imagine how the bridge workers felt when a train was rumbling across the bridge mere feet beneath them! Although unnerving for the workers, this was necessary so the railroad's service was not interrupted.

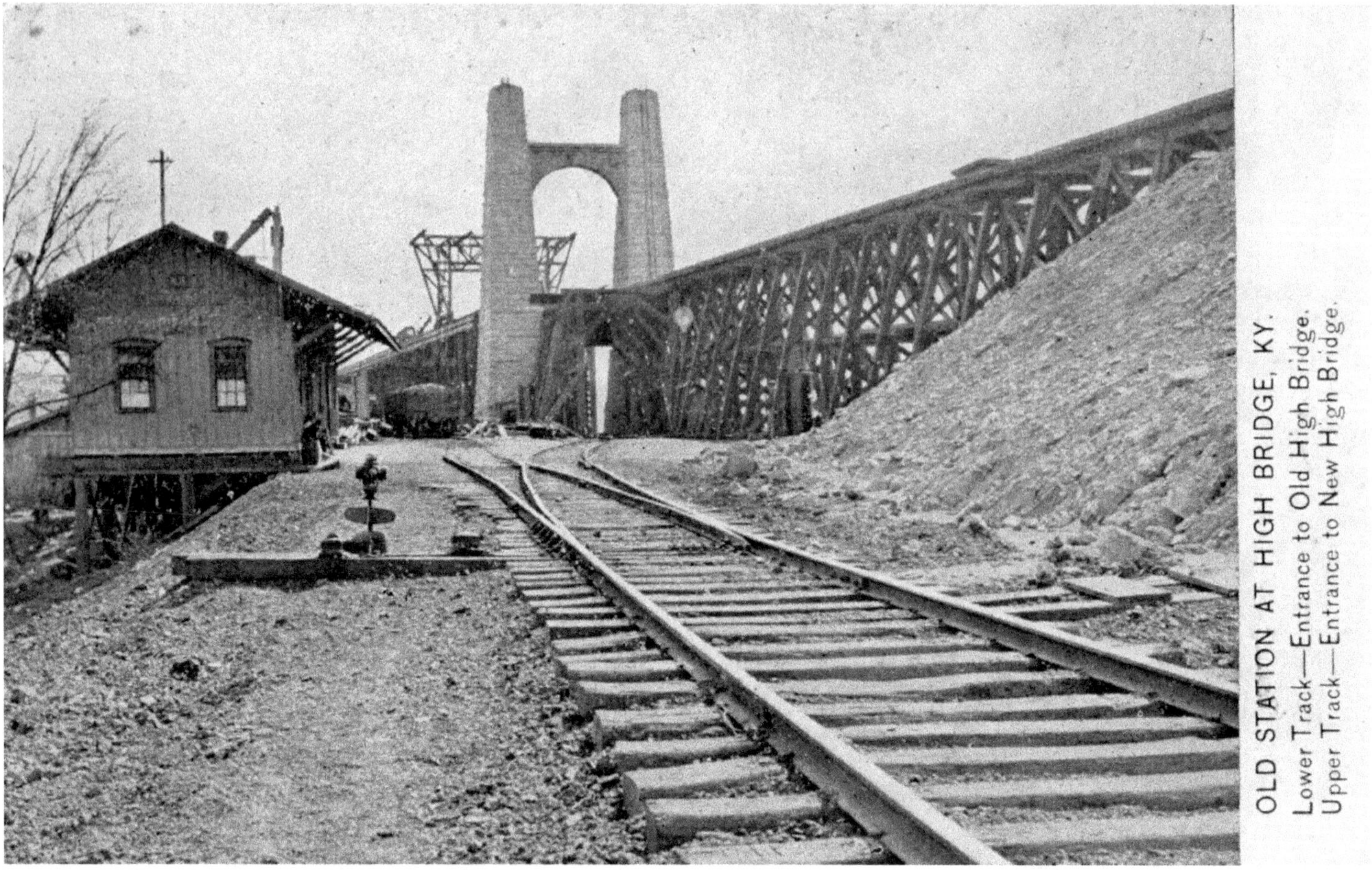

OLD STATION AT HIGH BRIDGE, KY.
Lower Track—Entrance to Old High Bridge.
Upper Track—Entrance to New High Bridge.

To accommodate the elevated track for the new bridge, a trestle was constructed and filled to add strength to the wooden structure. Note that the depot seen in this photograph is the original Victorian depot that once stood to the right of the bridge. Relocation became necessary because of its proximity to the tracks. When moved, the ornate Victorian woodwork was not reapplied to the depot.

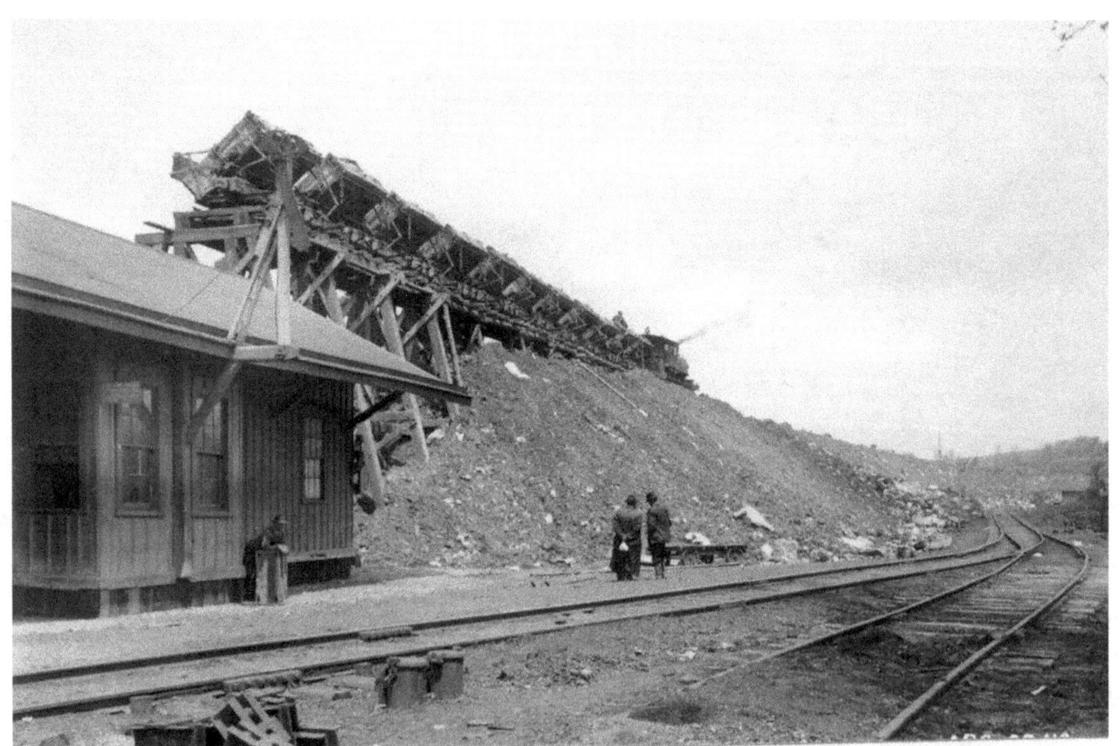

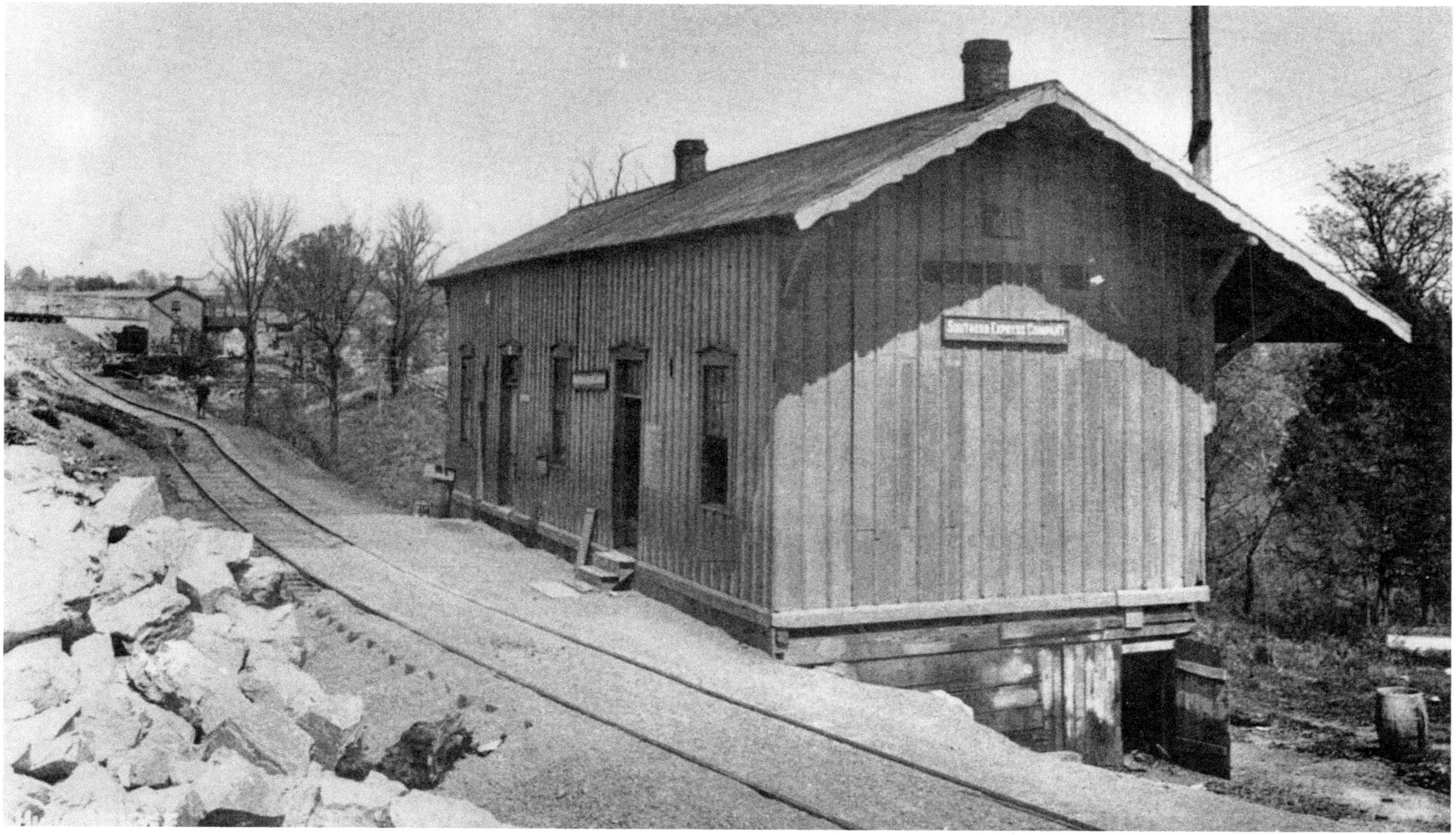

Relocated original depot with overhang removed.

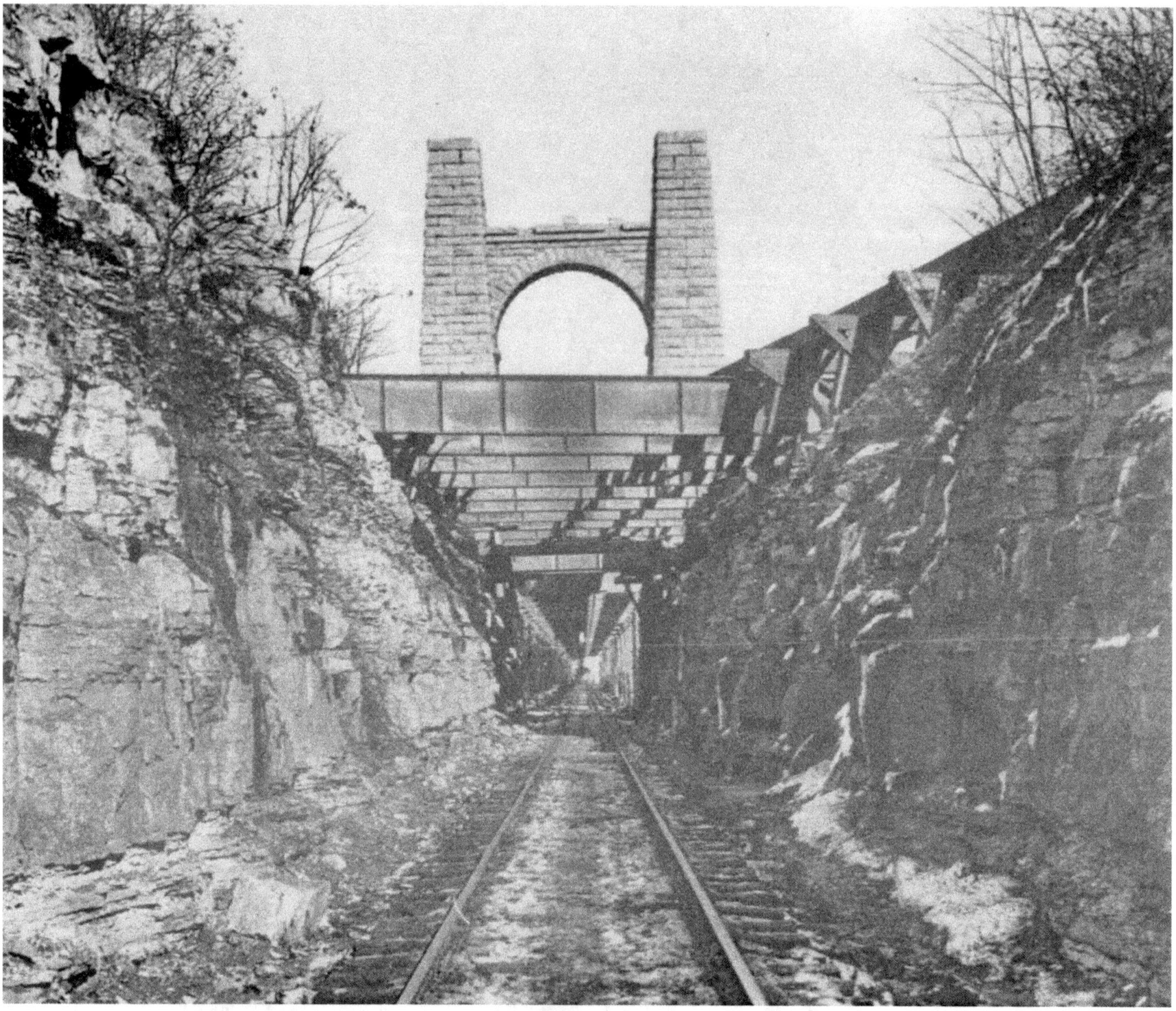

Approaching the bridge on the original track, while the track for the new bridge is being constructed above.

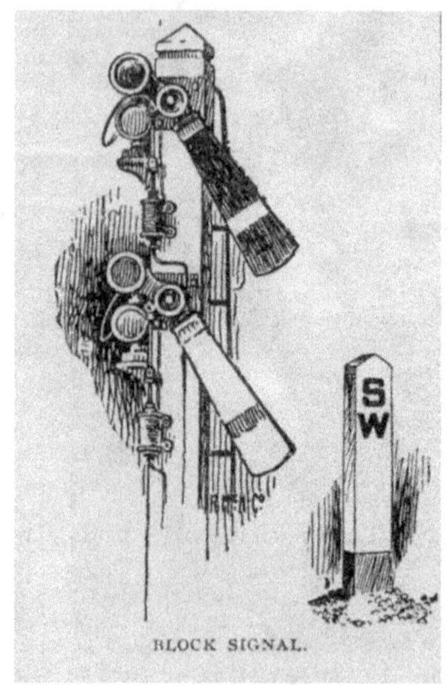

BLOCK SIGNAL.

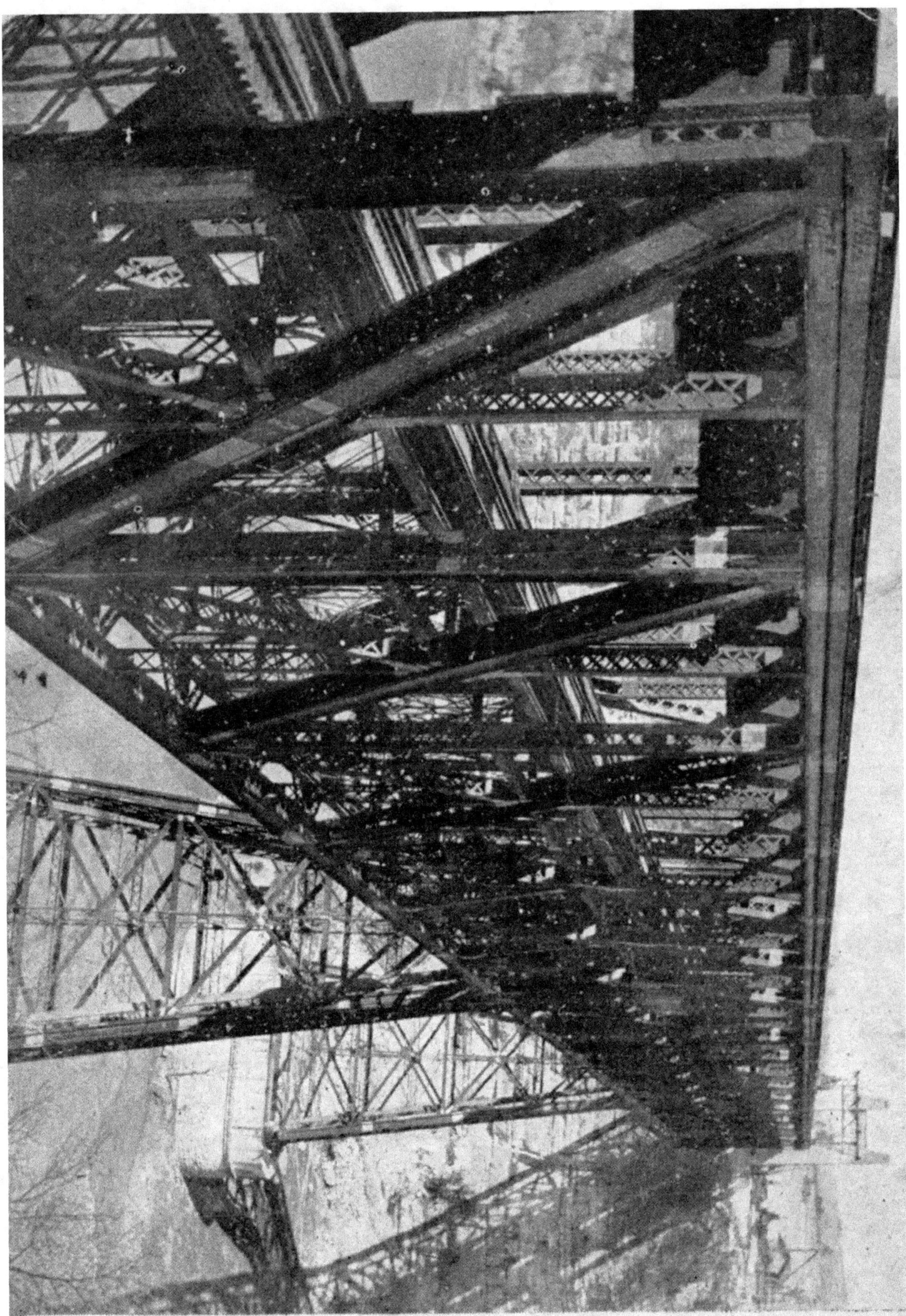

Detail View of Spans. Showing New Bridge built around old Bridge.

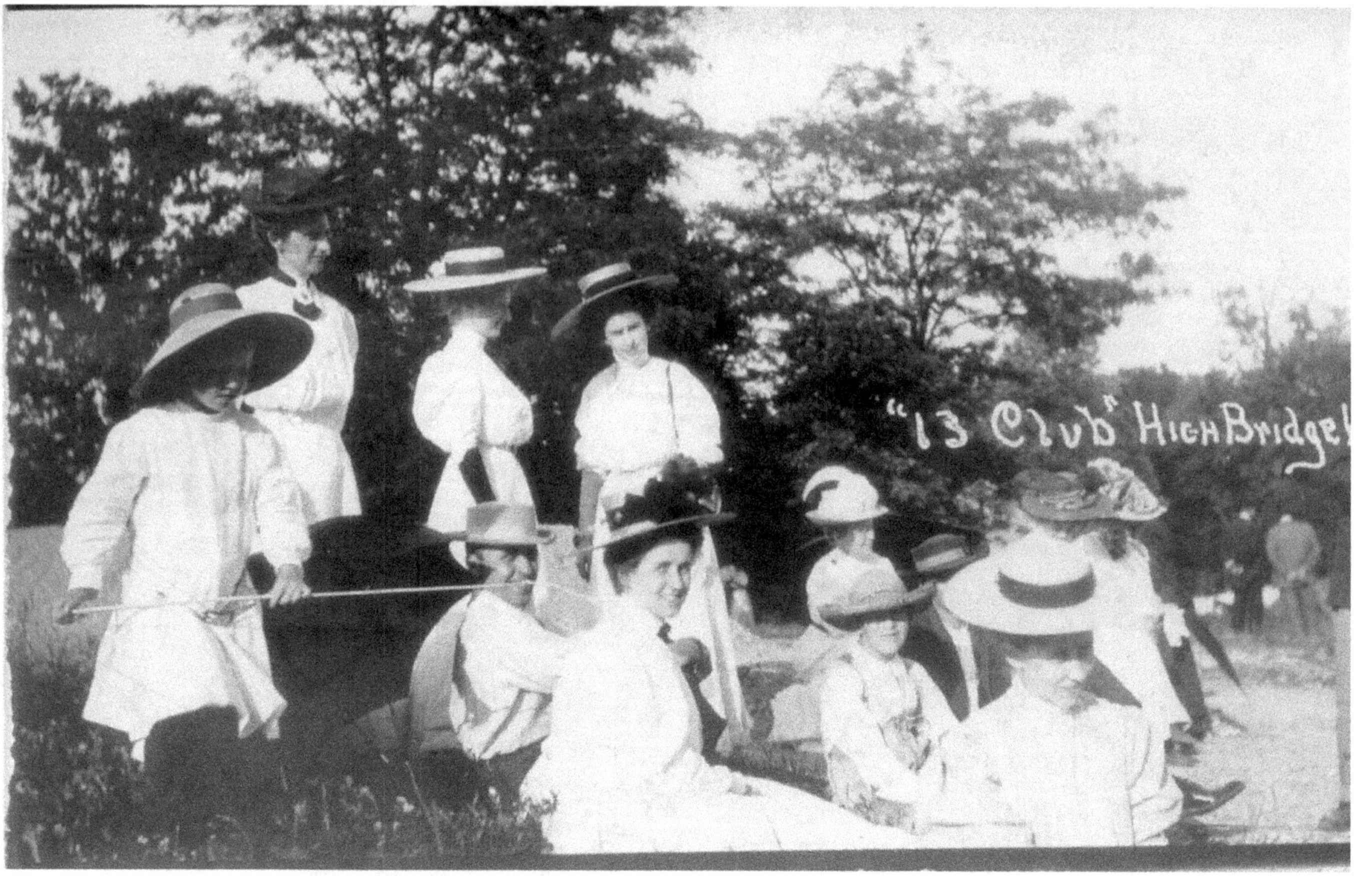

During the reconstruction of the bridge, architects, engineers and high level railroad officials, primarily from Cincinnati, would travel to High Bridge with their families to view the progress of the work. They would stay in boarding houses and vacant railroad section houses for periods of two weeks. While in High Bridge the group would engage in social activities and they referred to themselves as the "13 Club". Any meaning behind the club's name has been lost to time. .

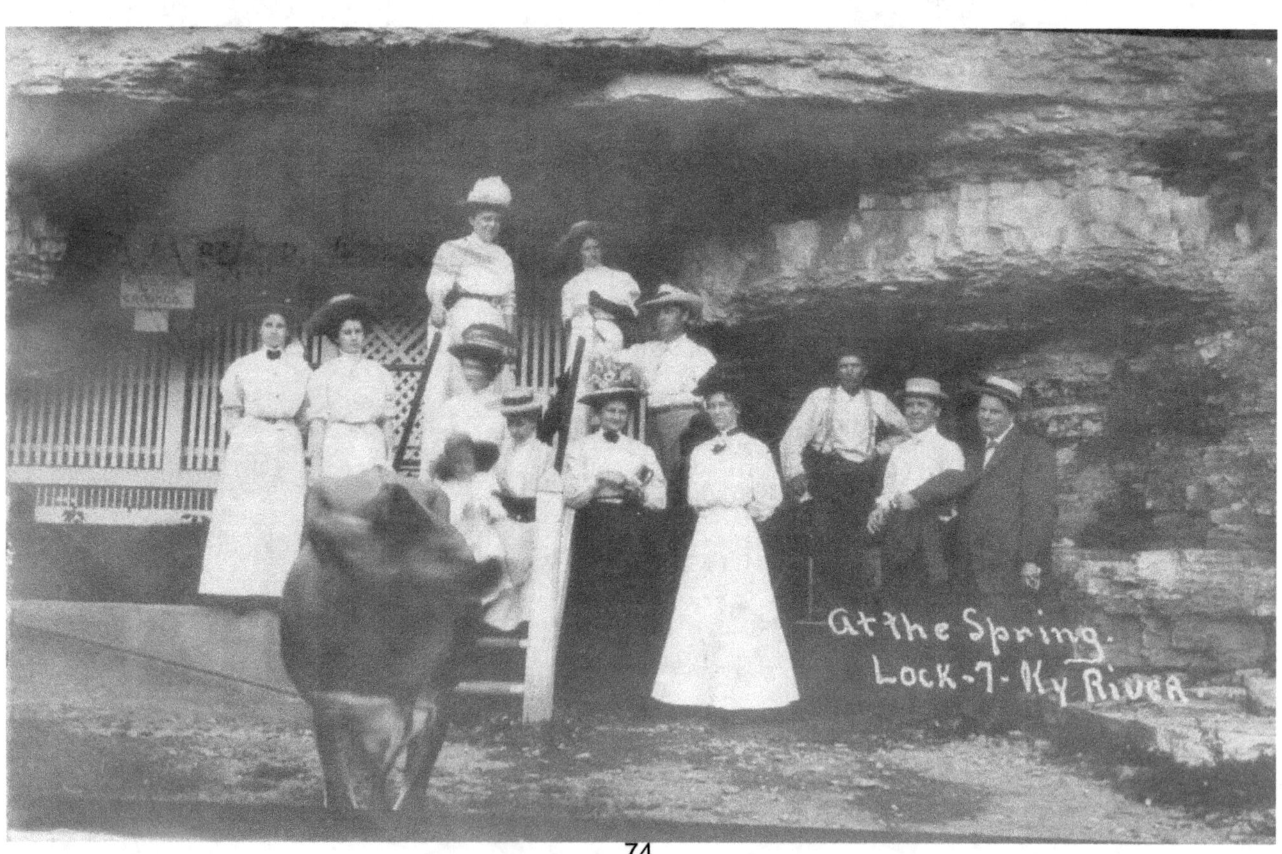

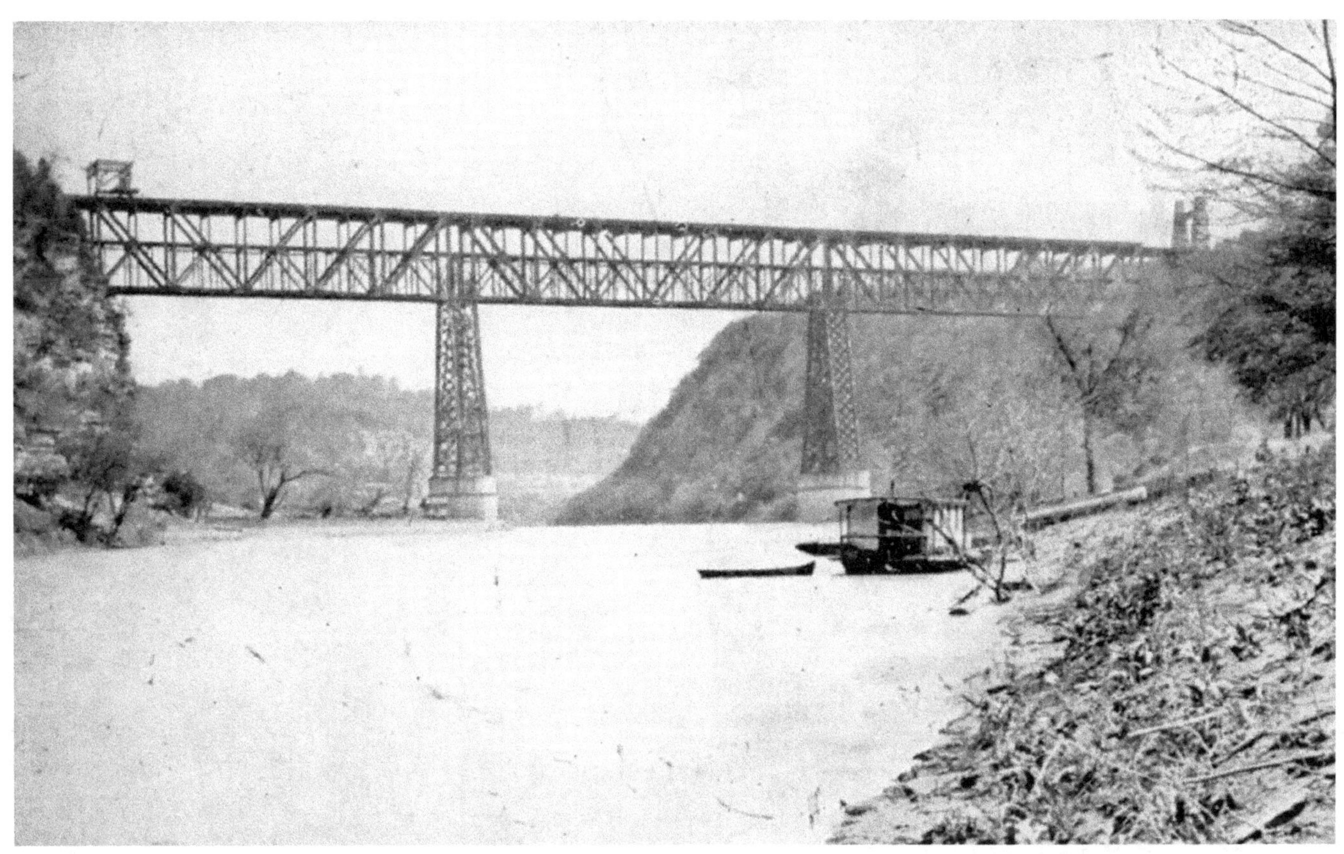

After more than a year, work on the new High Bridge was completed in September 1911 at a cost of $1,250,000. The first scheduled train to cross the bridge was on September 11, 1911, and the opening of the new bridge was greeted by as much fanfare as the opening of the original bridge.

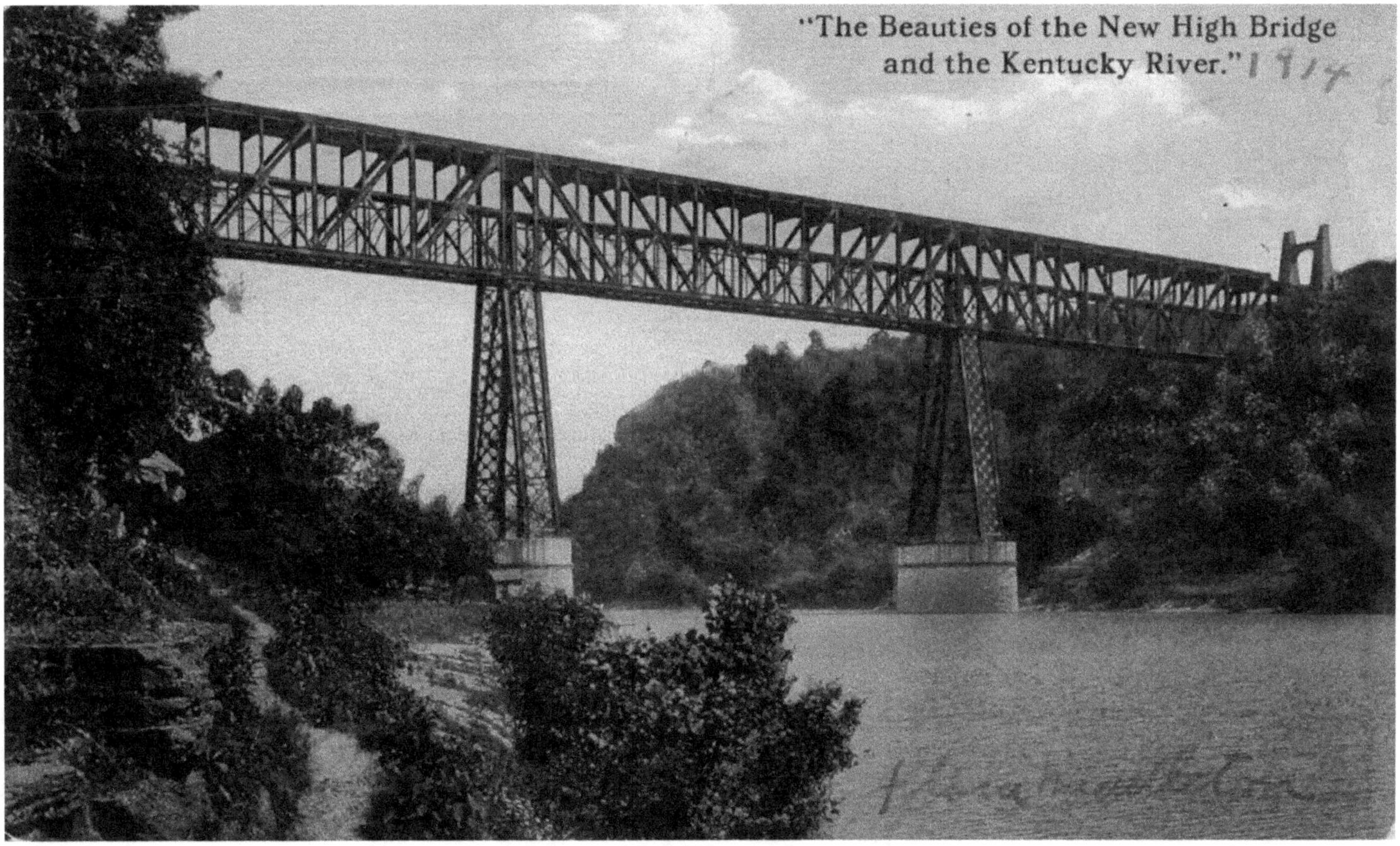

"The Beauties of the New High Bridge and the Kentucky River."

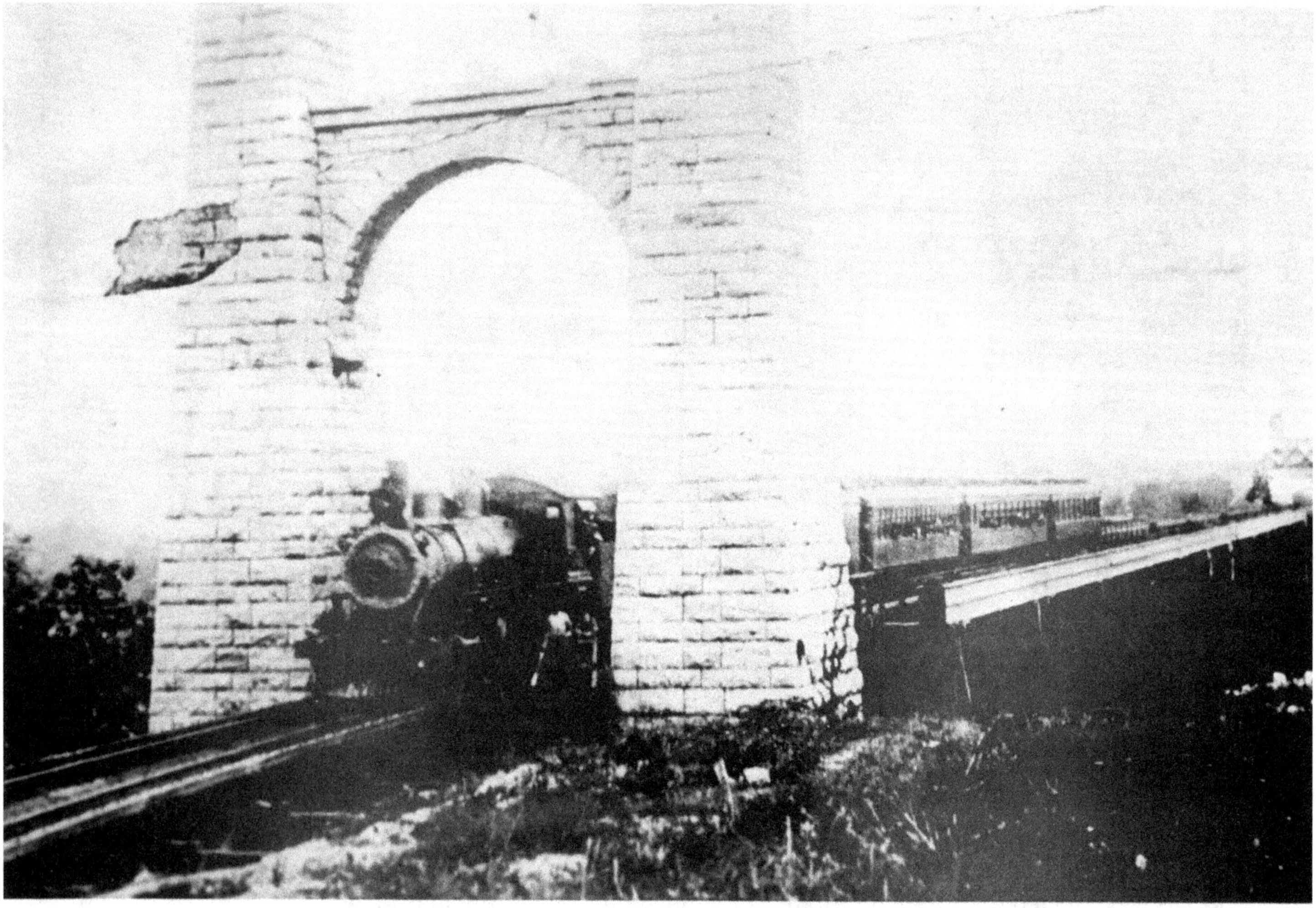

Heading south from Mercer County, locomotive engineer Clarence C. Horn took the first train across the bridge on September 11, 1911. Horn was chosen due to the fact that during his fourteen year career on the rails, he never had an accident.

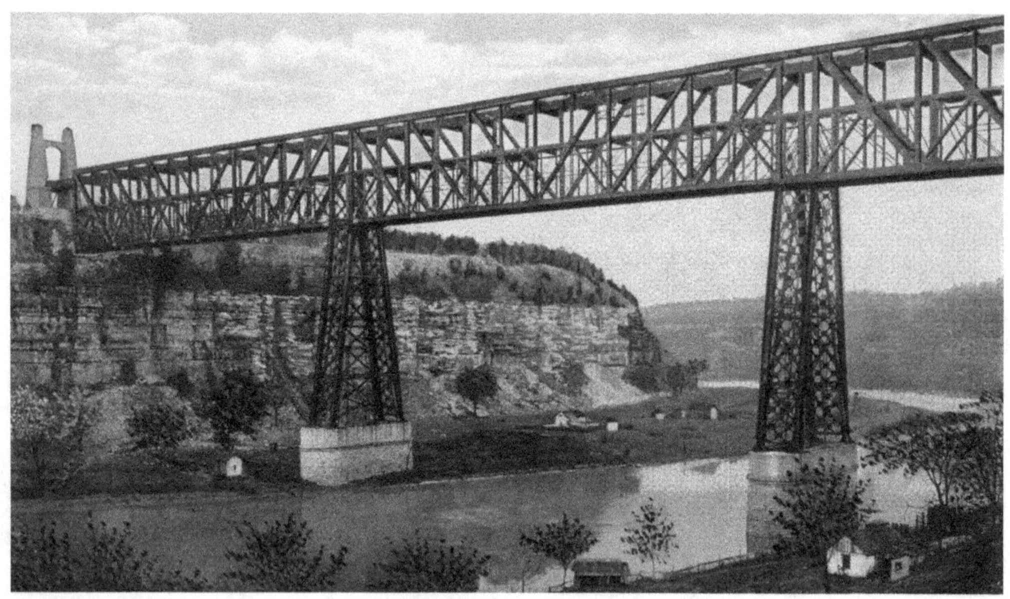

Not long after the new bridge opened, talks began on what to do with the portion of the original bridge that was to be removed from the interior of the new bridge. Judge George Kinkead proposed that the pieces be salvaged and then leased or sold to the Lexington Interurban Railway System. The pieces would be used to construct another bridge to carry Lexington's lighter, electric railroad. Kinkead's plan was rejected, and the original bridge was sold for scrap.

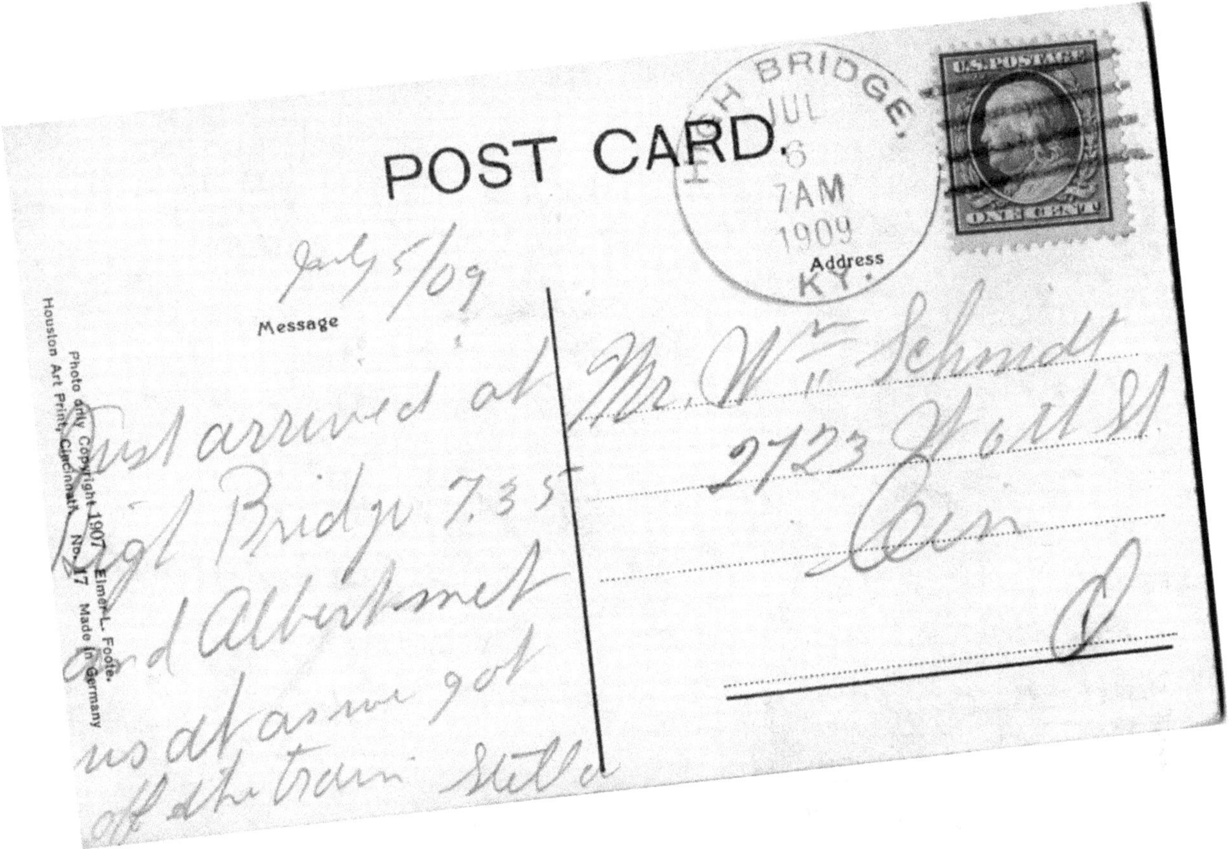

Although many pieces of correspondence passed through the High Bridge Post Office until its closing in 1976, finding a surviving postcard with a High Bridge postmark is very rare today. Prior to postmarking mail as being from High Bridge, the postmark would read Kentucky River Bridge, obviously during the bridge's construction, and North Towers before the community was renamed High Bridge around the time of the bridge's dedication.

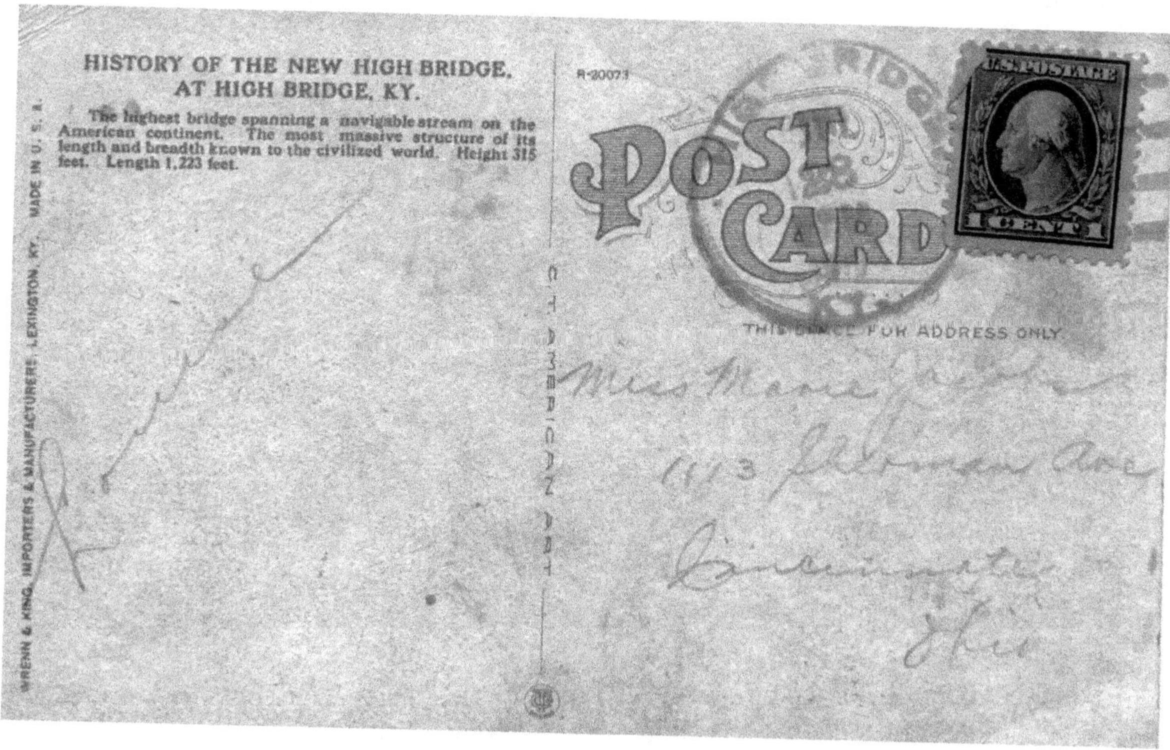

Dix & Kentucky Rivers near High Bridge, Ky.

Postcards offer a cameo of daily life at the turn of the century. Part of this postcard postmarked December 9, 1909, reads as follows: "I am at the office and the boss thinks I am hard at work of course, it is not good for the boss to know everything. Tell Selena she had better practice euchre a little more for you and I are going to put it on her and Barry the next time we play."

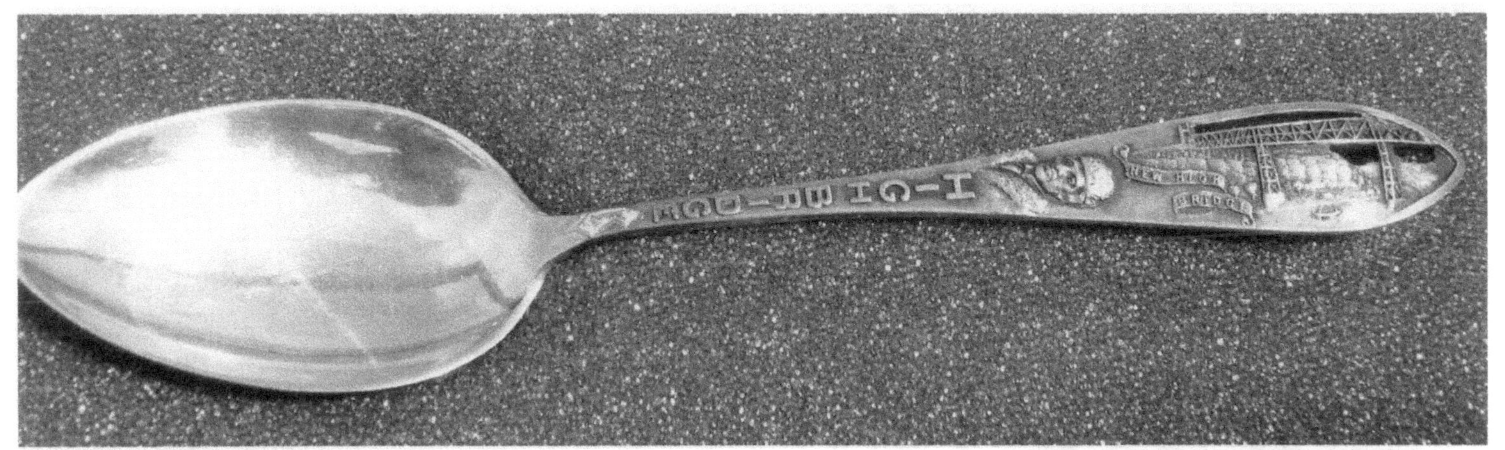

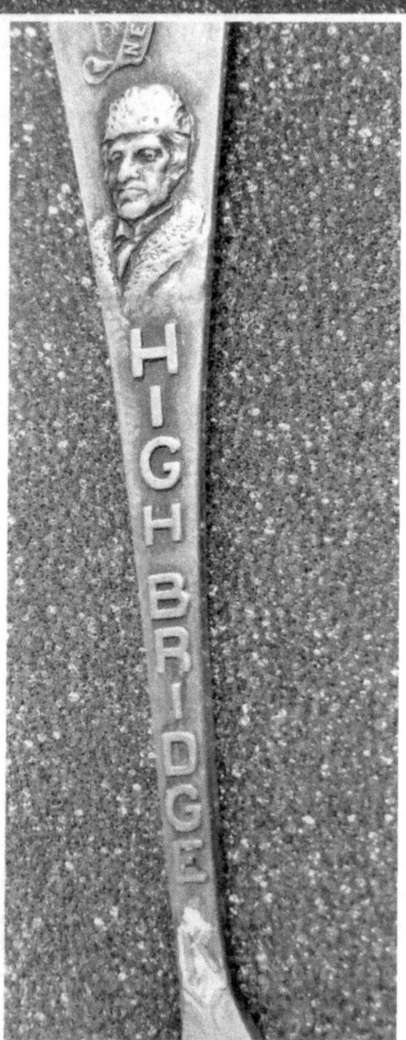

In what had to be one of the most elegant souvenirs of High Bridge, an October 9, 1914, newspaper advertisement for this spoon read as follows: “Mrs. Mattie Dorman McKee has designed and is selling a very beautiful and useful souvenir of High Bridge in the shape of a tea spoon. It is of oxidized sterling silver with plain bowl, but the handle is very beautifully decorated with a bust of Daniel Boone and a splendid etching of the new bridge with New High Bridge in raised letters. The design is truly a work of art and all who care for an inexpensive and useful souvenir can write to Mrs. McKee at High Bridge. The spoon has only been out a few weeks and Mrs. McKee has already received many orders.” Mrs. McKee’s initials appeared on the reverse side of the spoon’s bowl.

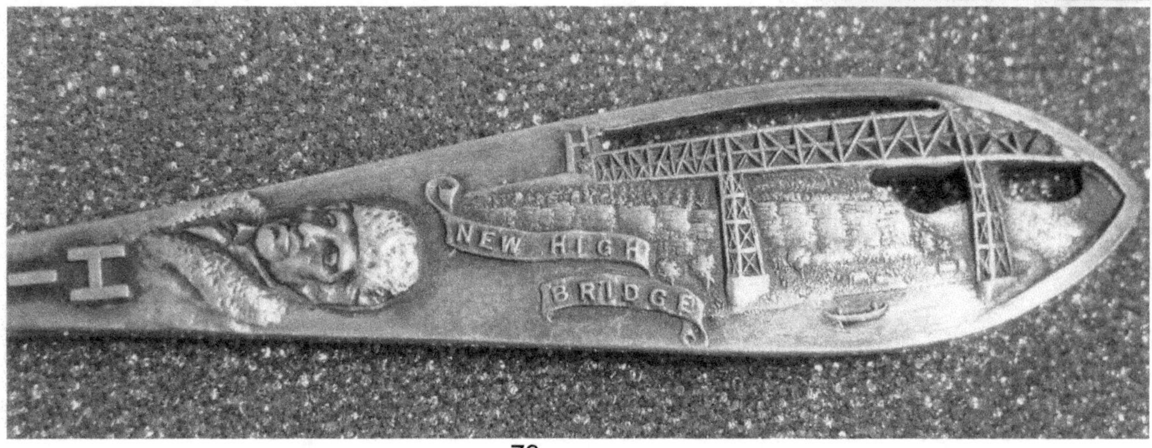

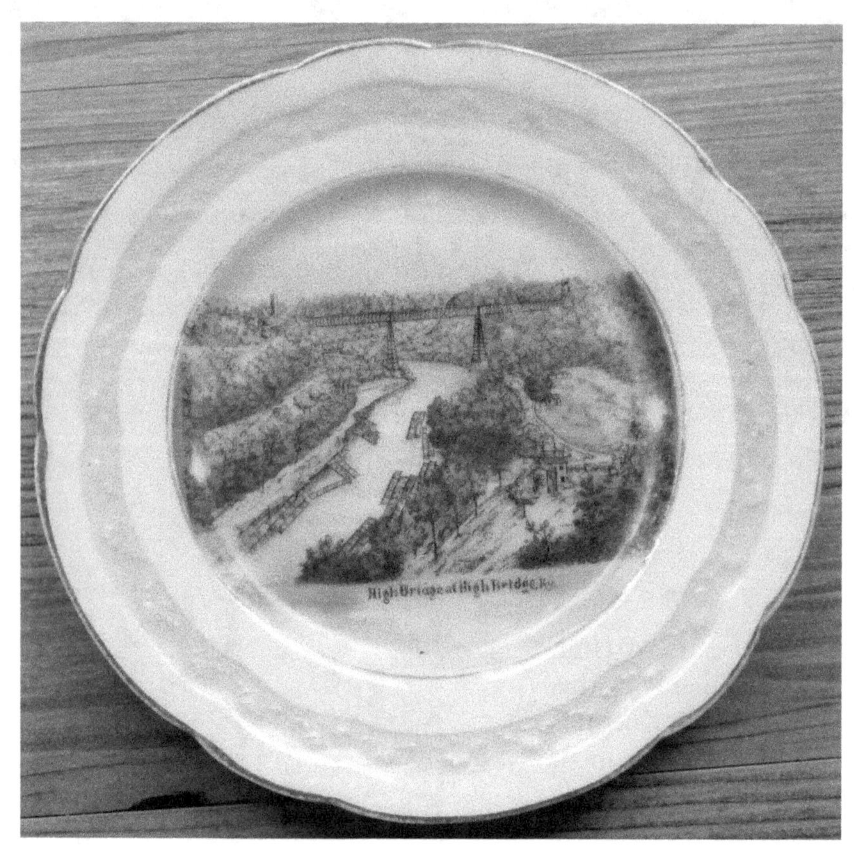

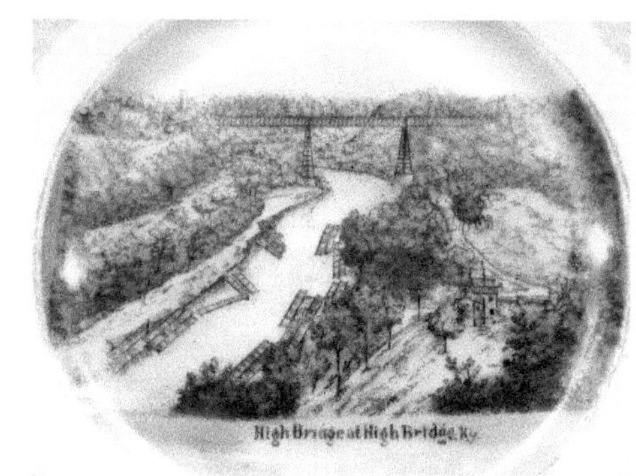

Detail view of Dresden China souvenir plate.

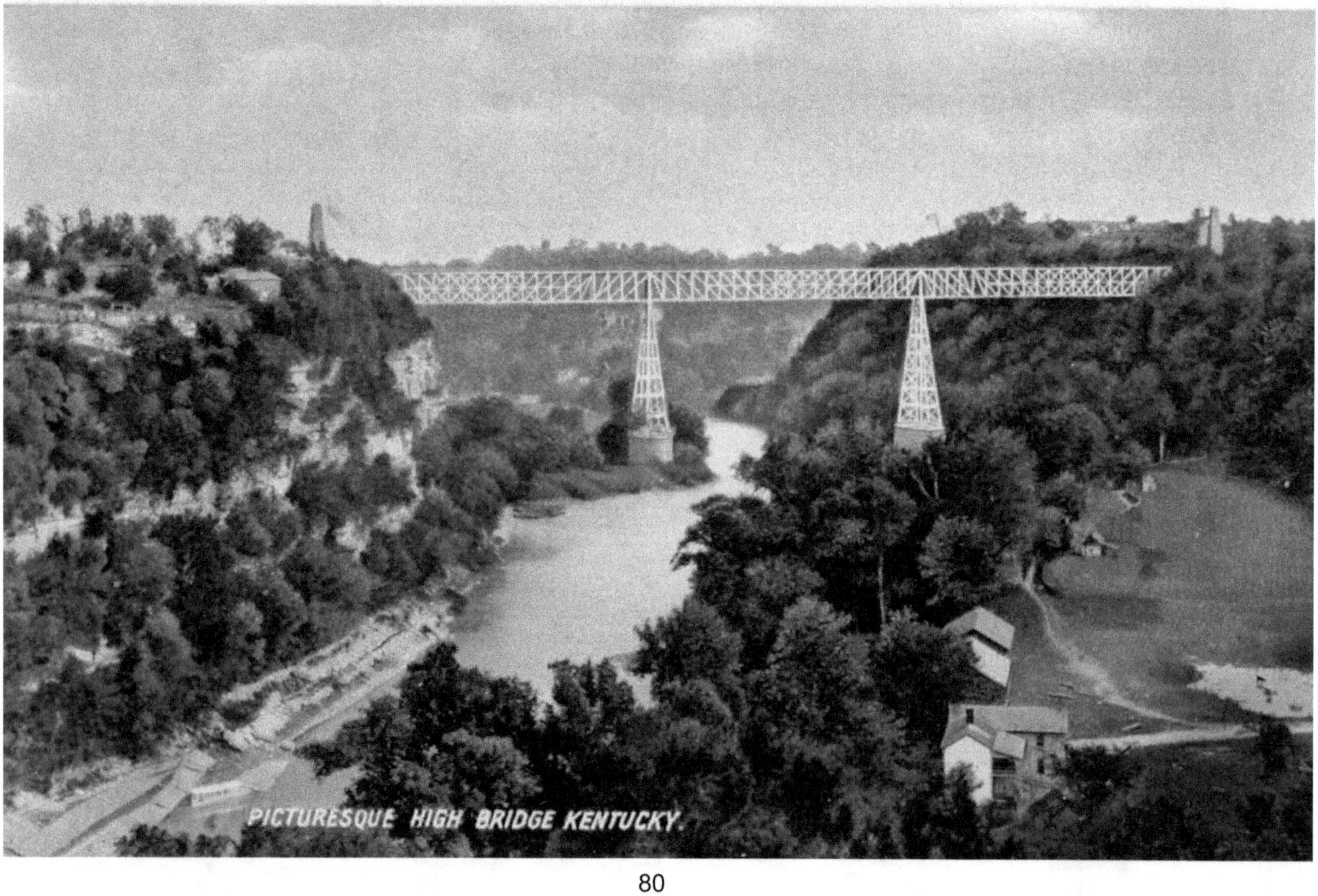

PICTURESQUE HIGH BRIDGE KENTUCKY.

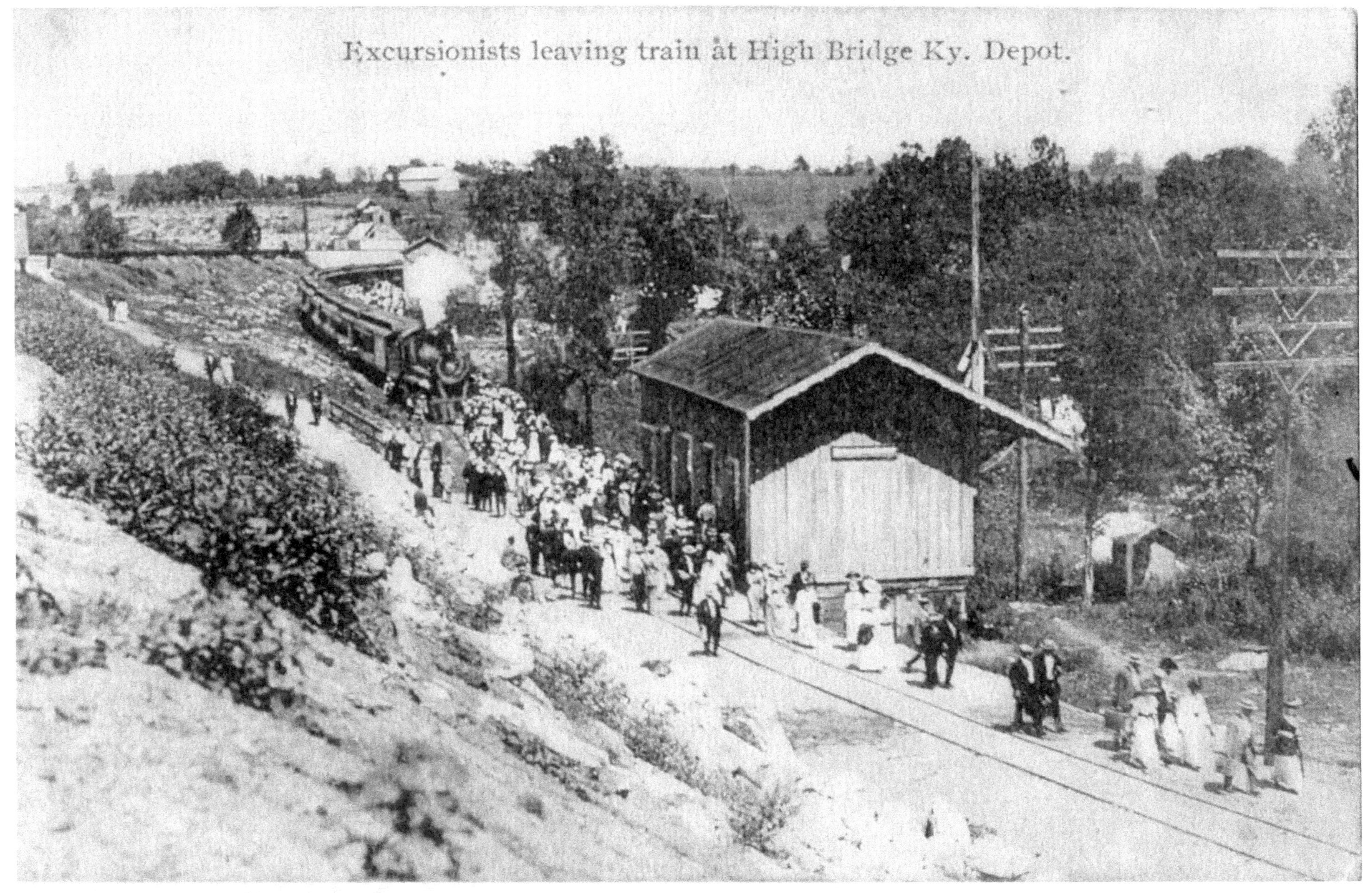

Taken from the elevation of the new bridge still under construction, this photograph of an excursion train with passengers debarking was taken circa 1910. The spur track was created for trains to pull off for passengers. It is still in existence as Old Saw Mill Road. However, the tracks have been removed, and the roadway only serves automobile traffic.

The second High Bridge depot was built farther down the tracks and stood at the higher level of the new bridge. The uninviting and plain structure lacked the warmth and ornate exterior trims of the original depot that had welcomed travelers for many years.

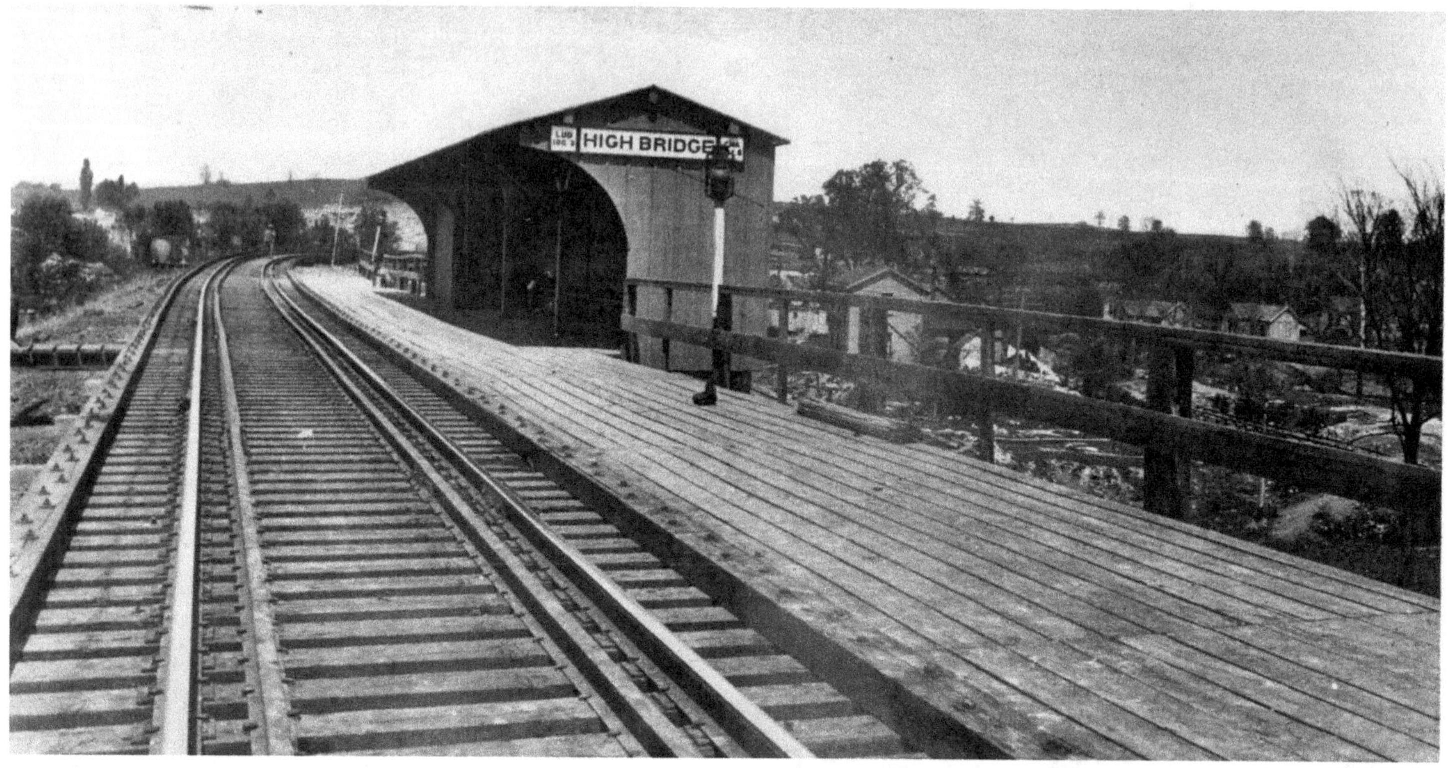

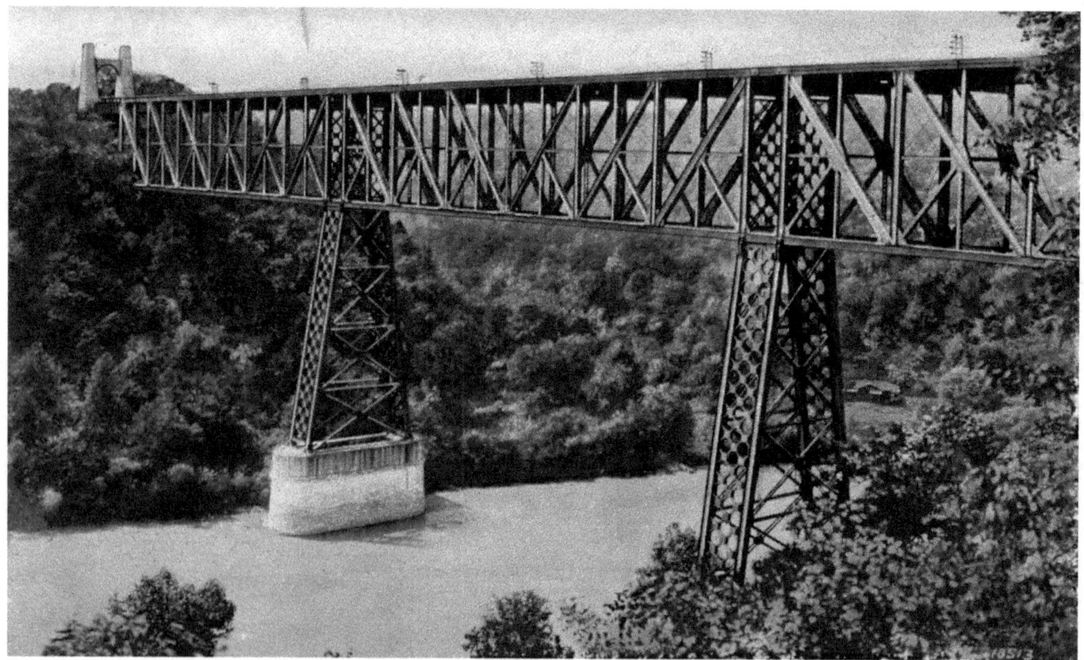

The original bridge and the new bridge were always painted every few years. But because of the massive size of the new structure that practice was stopped around 1915. The project was considered too expensive and too dangerous for workers because of the amount of rail traffic on the bridge and its height.

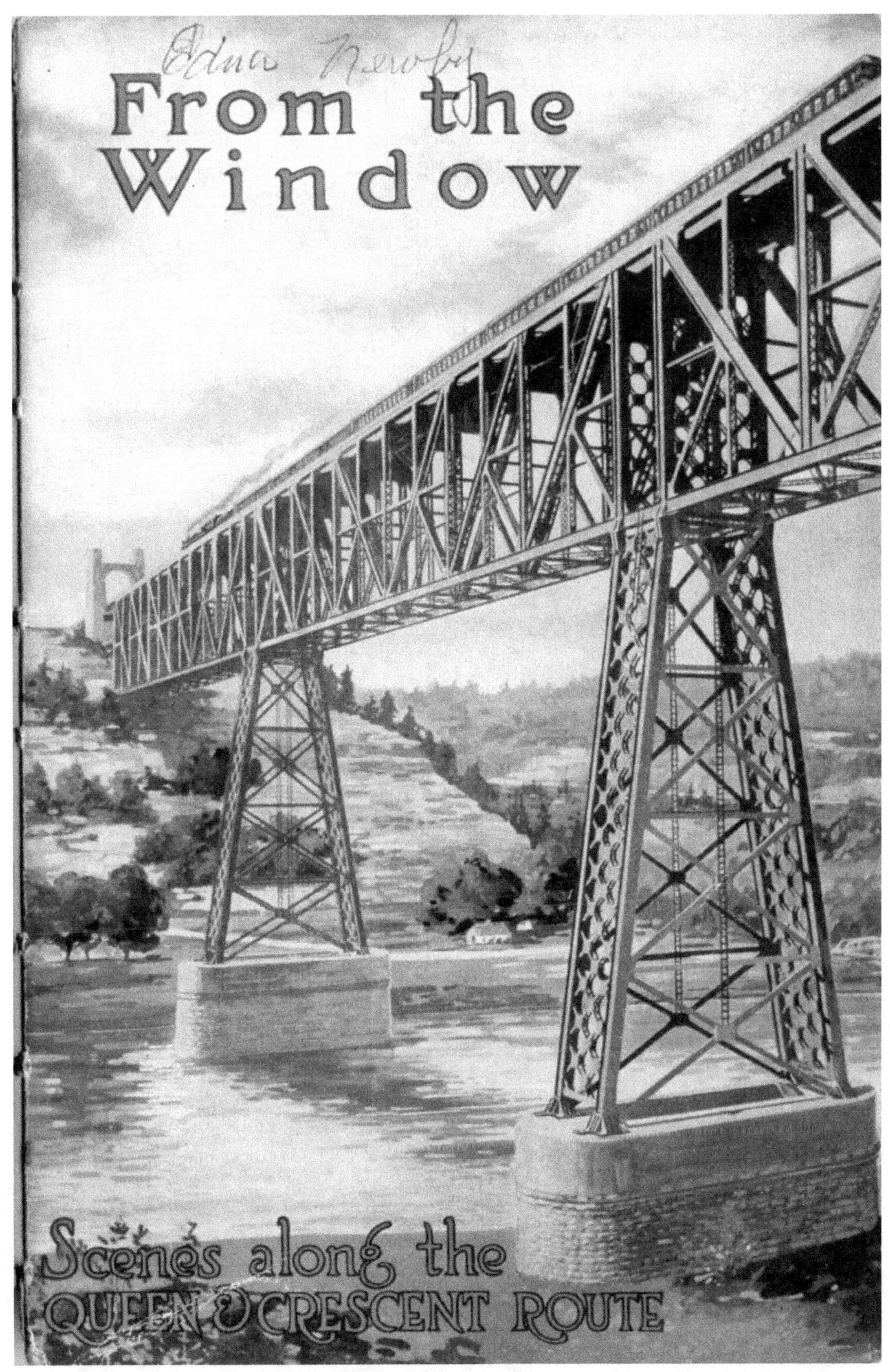

This drawing of High Bridge appeared in an April 1912 *From the Window* book. Given to passengers of the Queen and Crescent Route special train between Cincinnati, Ohio, and Chattanooga, Tennessee, the book contained many photographs and a brief bit of information about sights that would be seen from the window by passengers on the train.

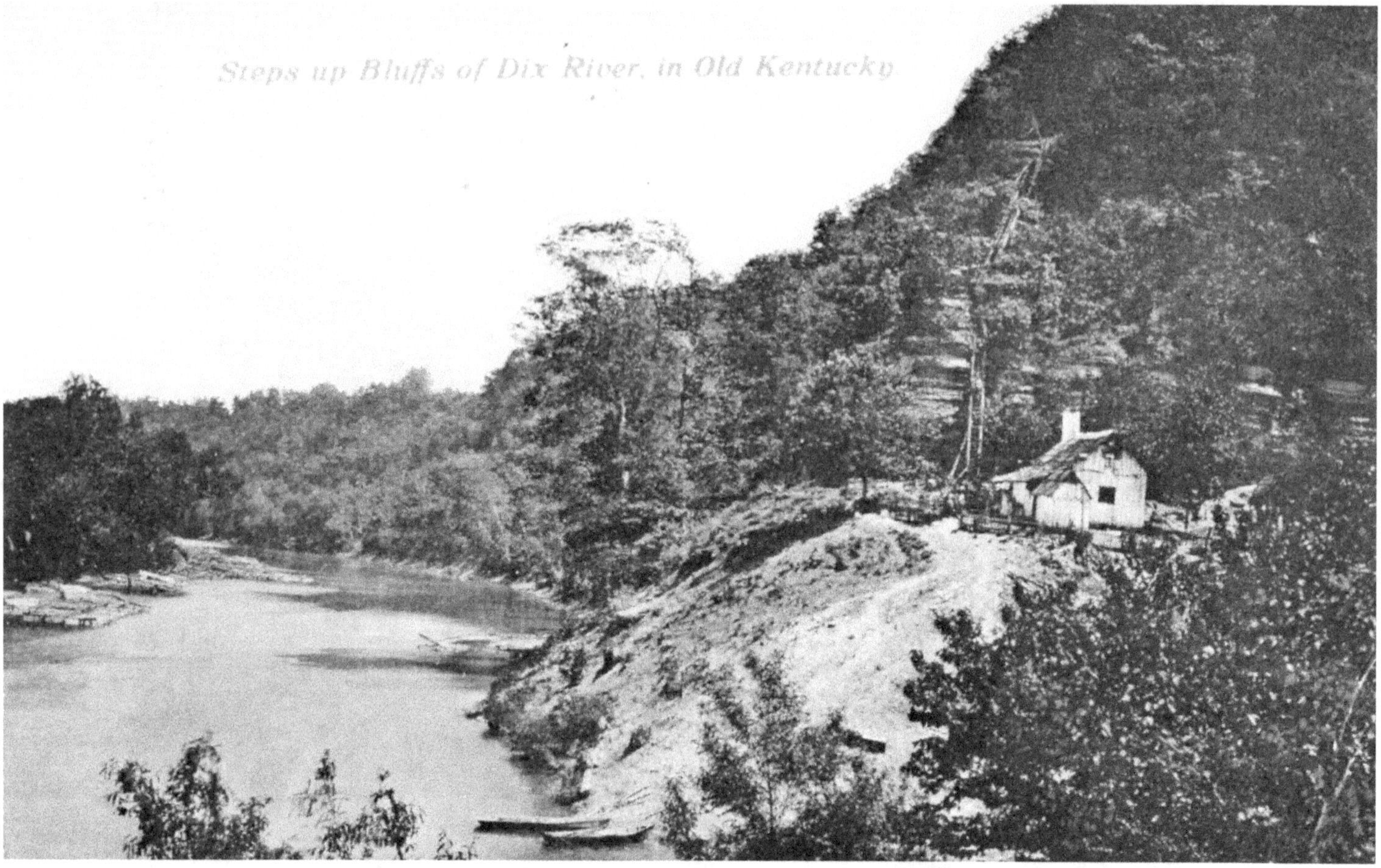

Steps up Bluffs of Dix River, in Old Kentucky

The popularity of the stairs down the cliff from High Bridge Park led other residents to do the same. One resident, Wash Hicks, built a small ferry landing on the edge of his property at the mouth of the Dix River. He would bring paying passengers on his boat to stairs he had built up the cliff to visit a cave in the cliffs.

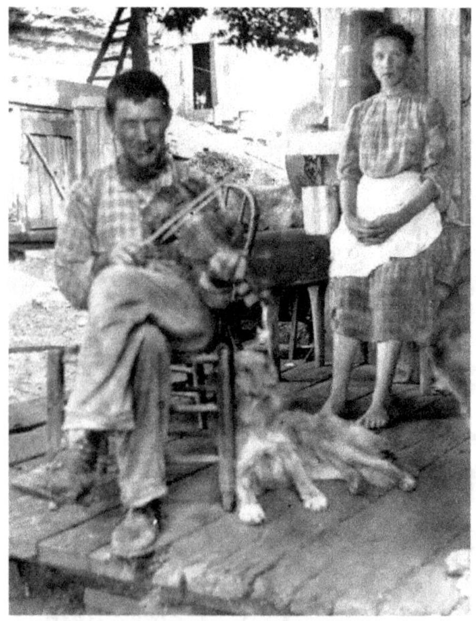

Wash and Amanda Hicks.

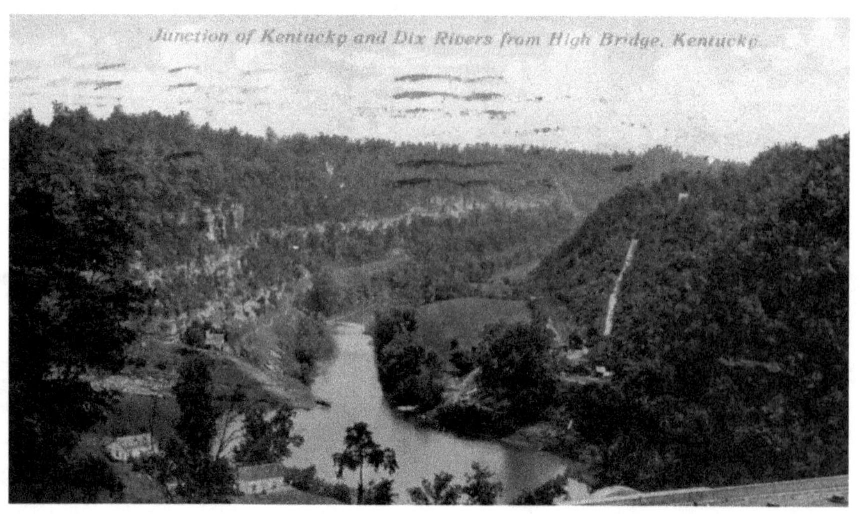

Excursion group in cave near High Bridge.

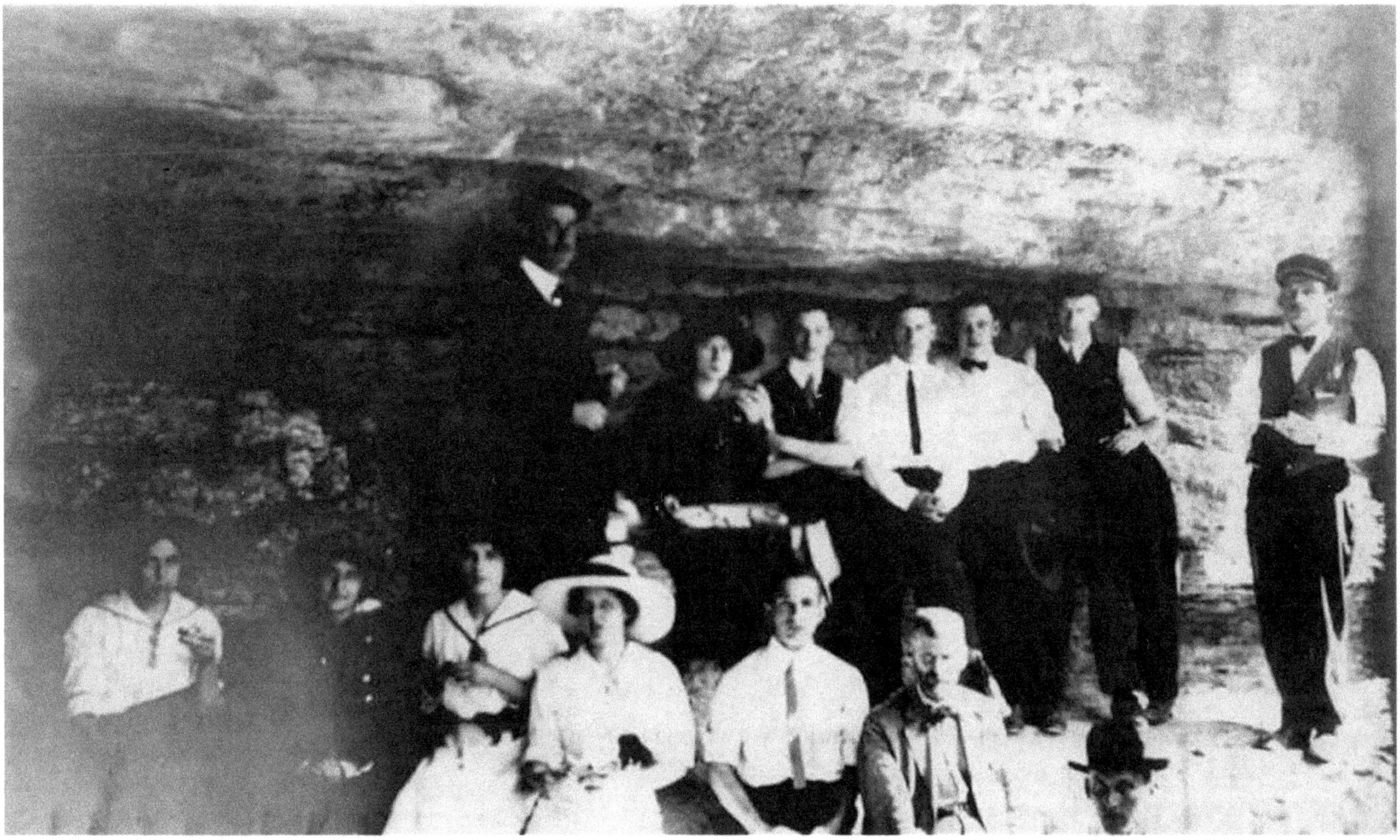

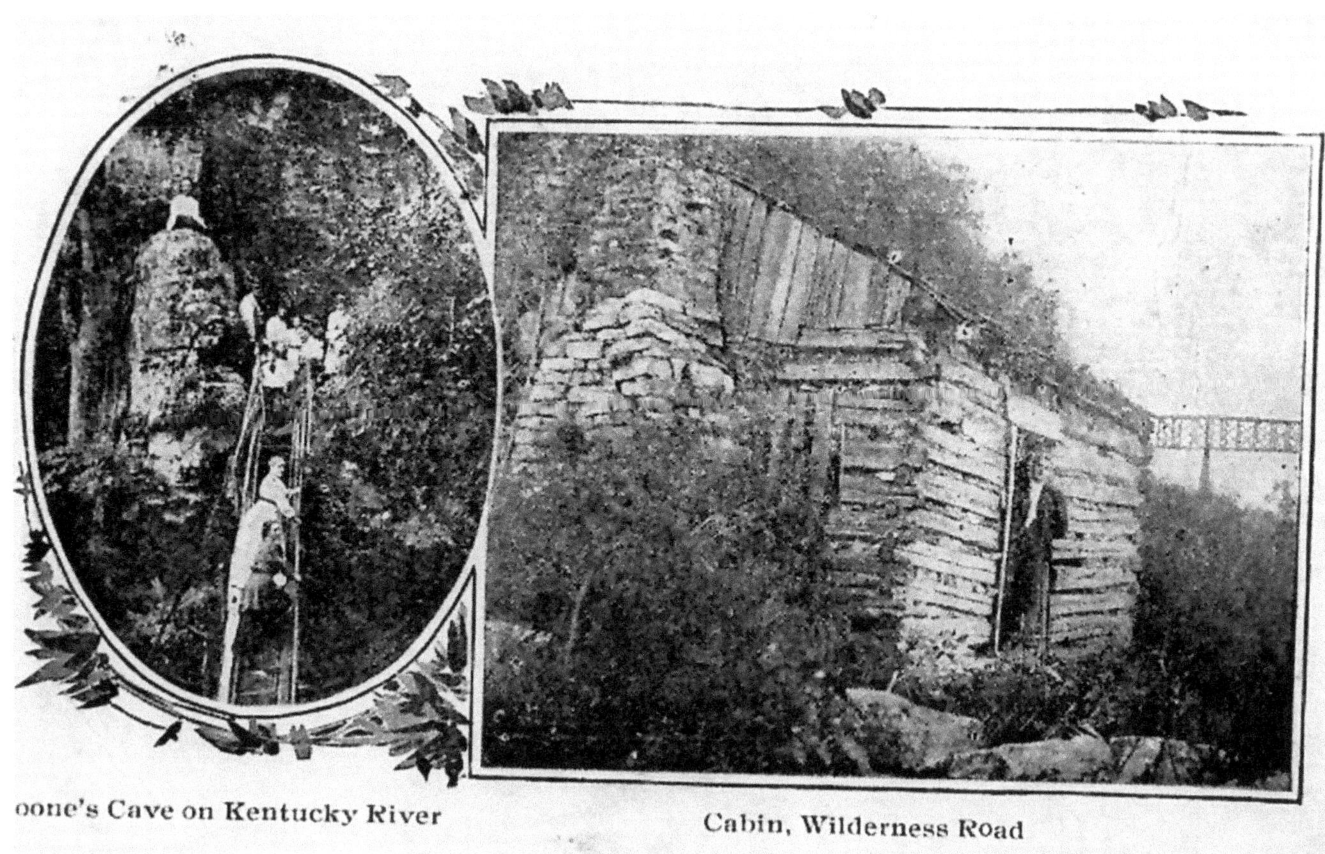

oone's Cave on Kentucky River

Cabin, Wilderness Road

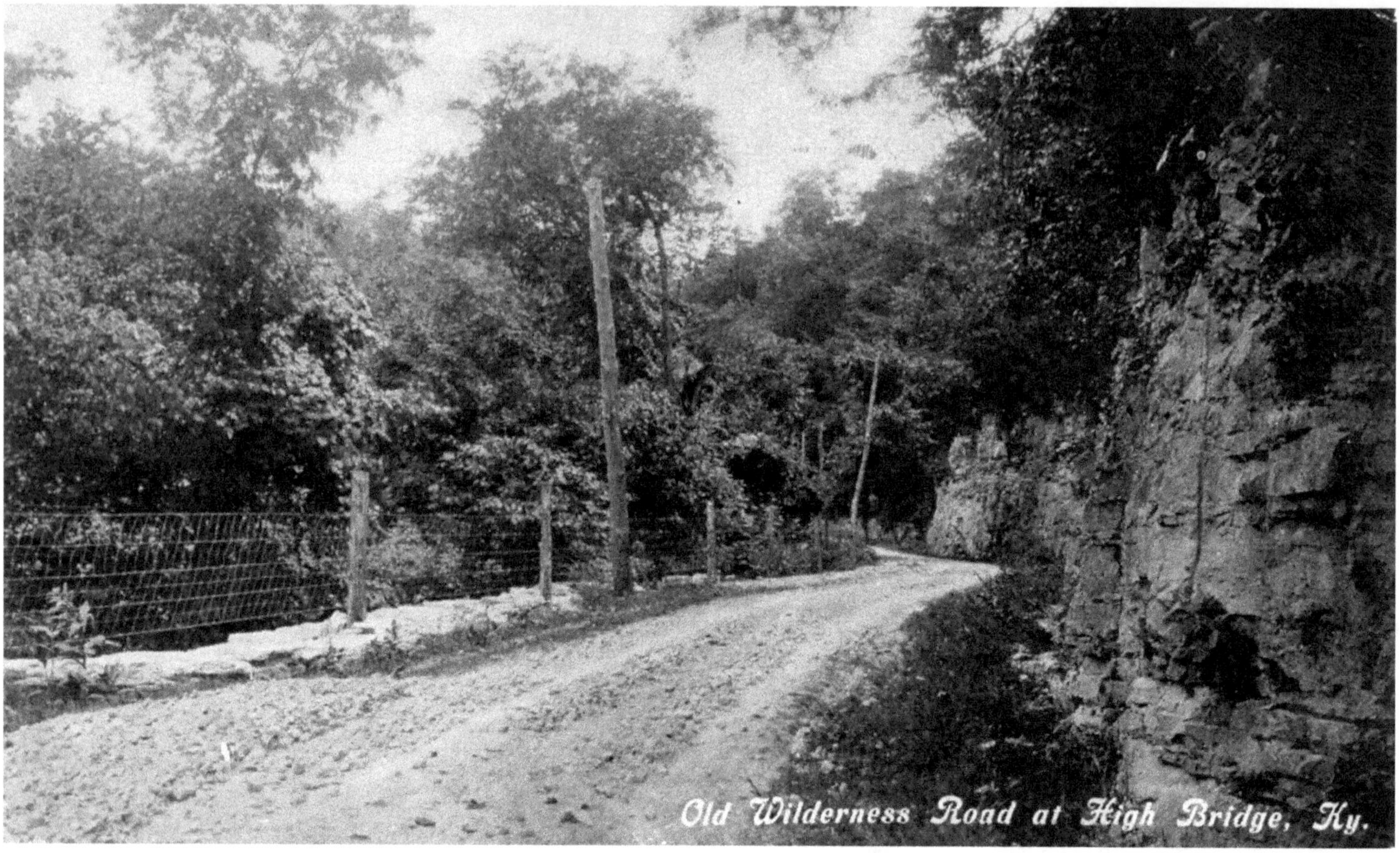

Old Wilderness Road at High Bridge, Ky.

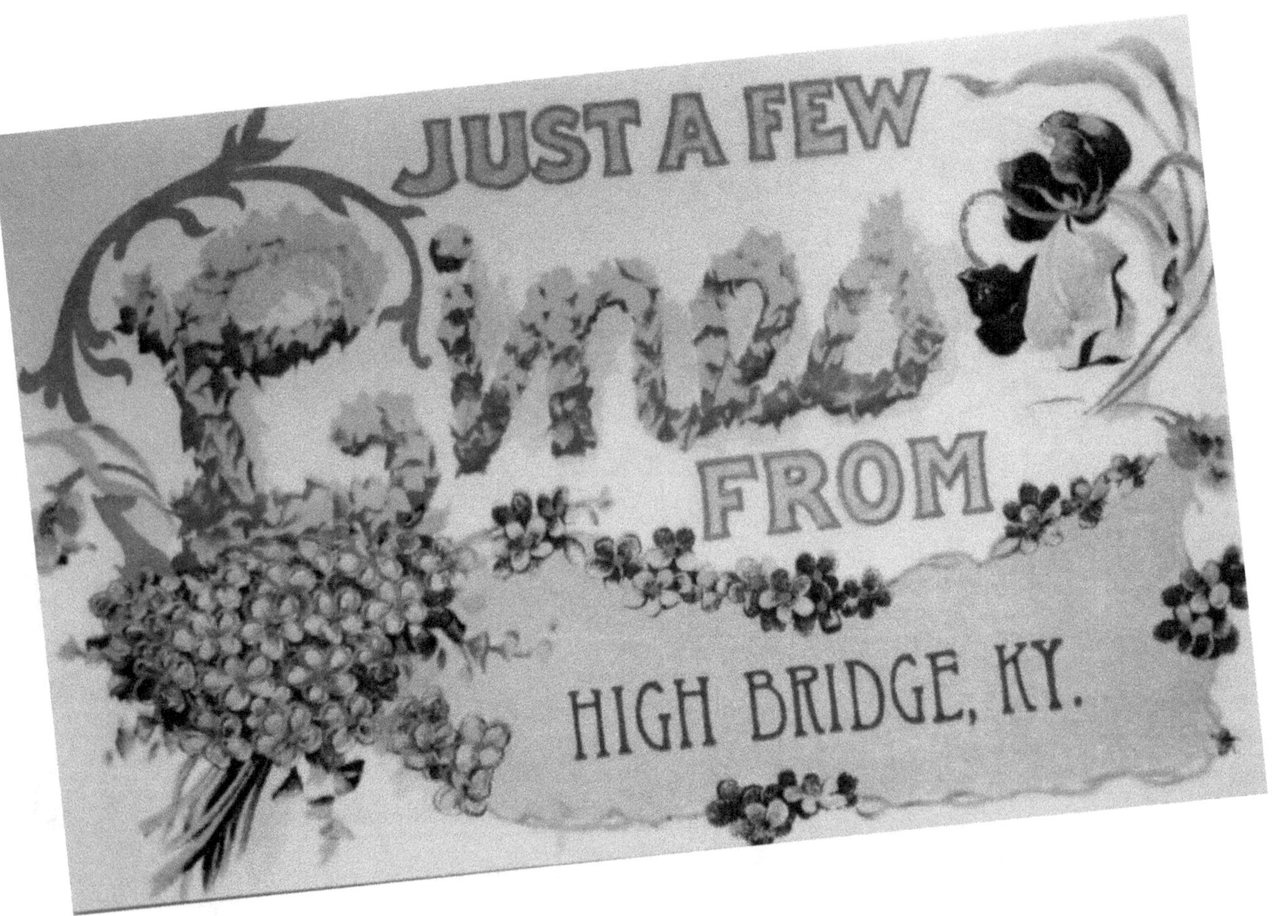

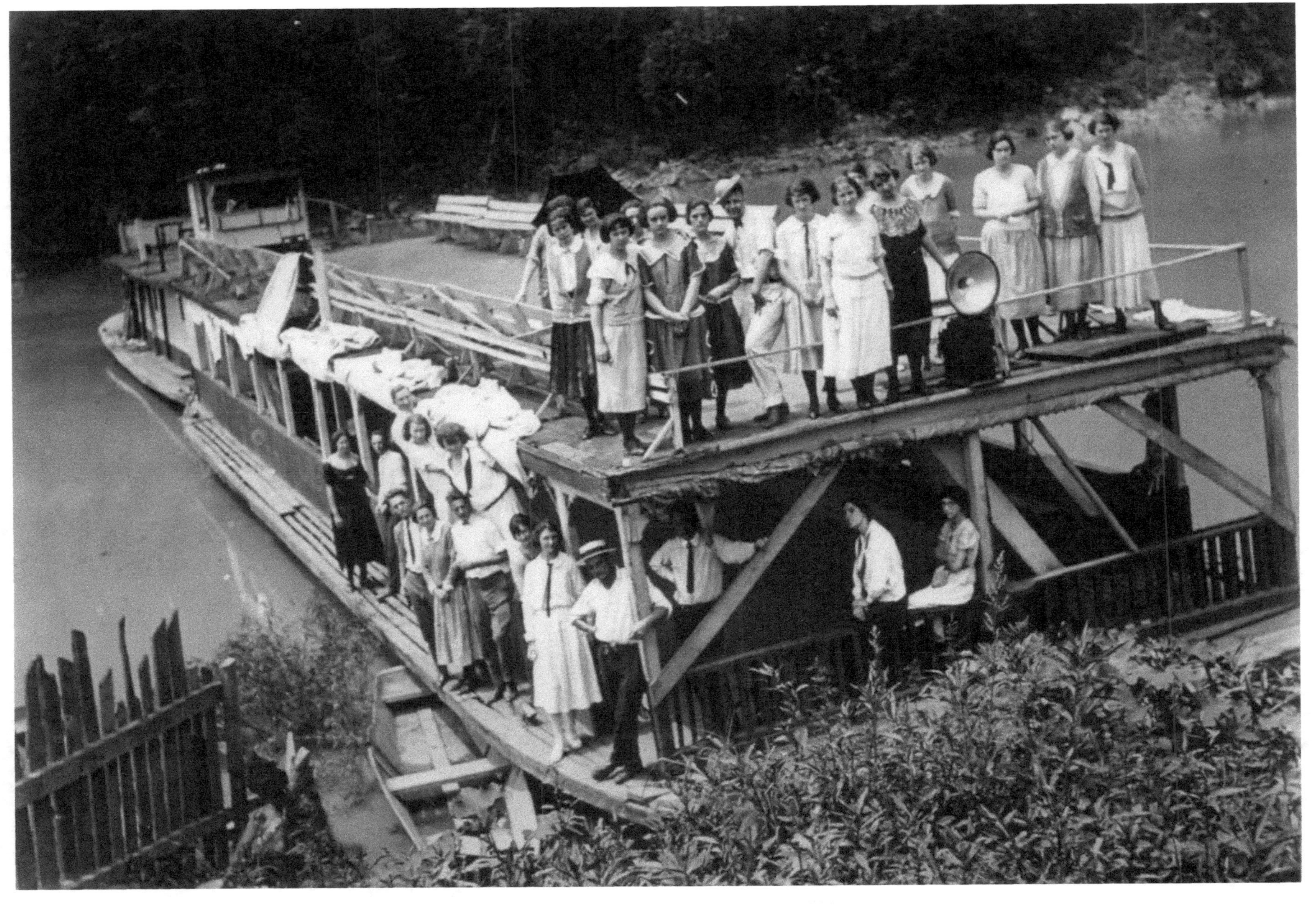

The *Chester* was one of the many locally constructed and owned excursion vessels made available for people who came to High Bridge by train, automobile, or steamboat to visit the area. Boats such as the *Chester* would often take passengers from the Shaker landing to the mouth of the Dix River and back. This photograph dates from 1922.

High Bridge Park

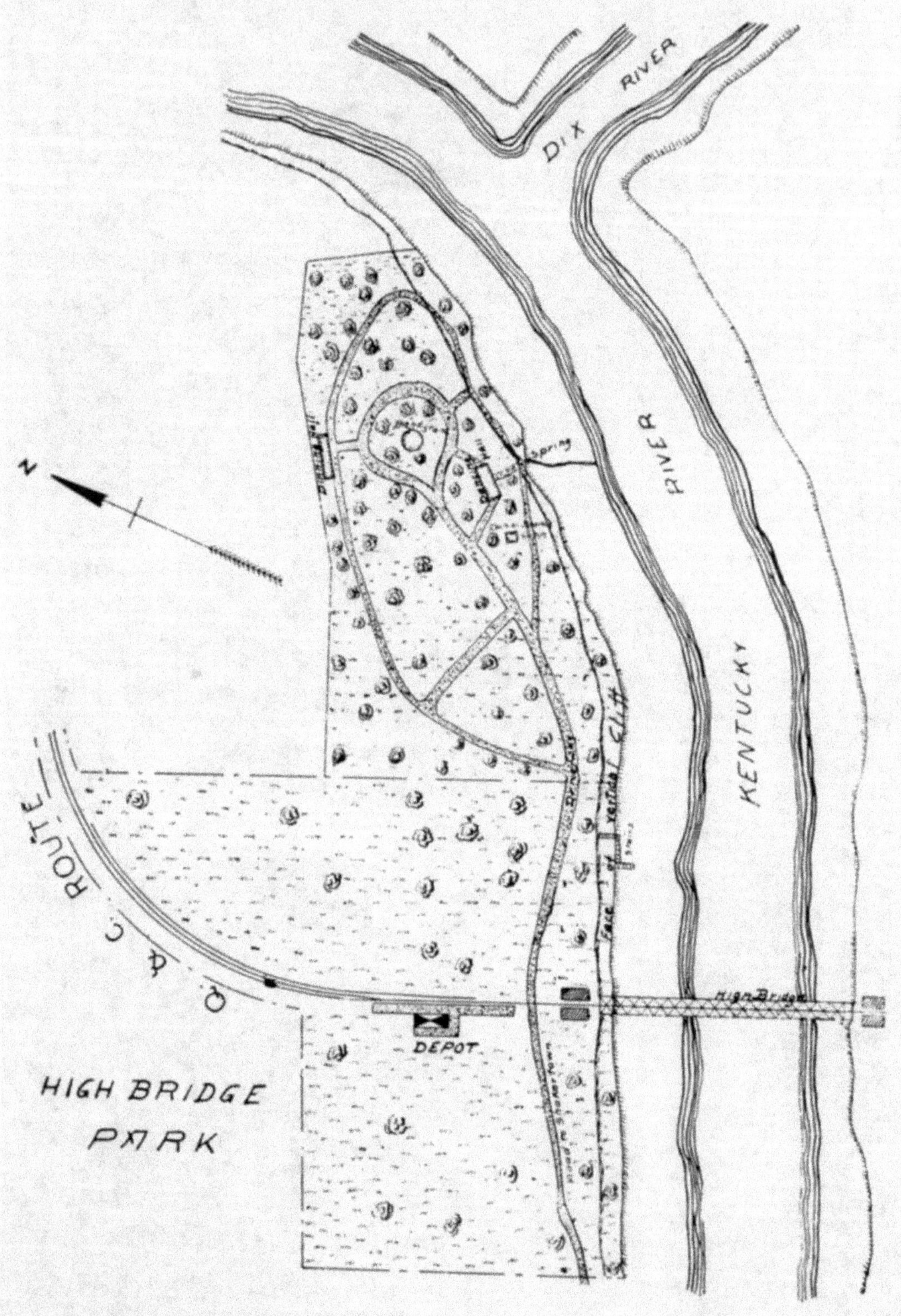

At

High Bridge Kentucky.

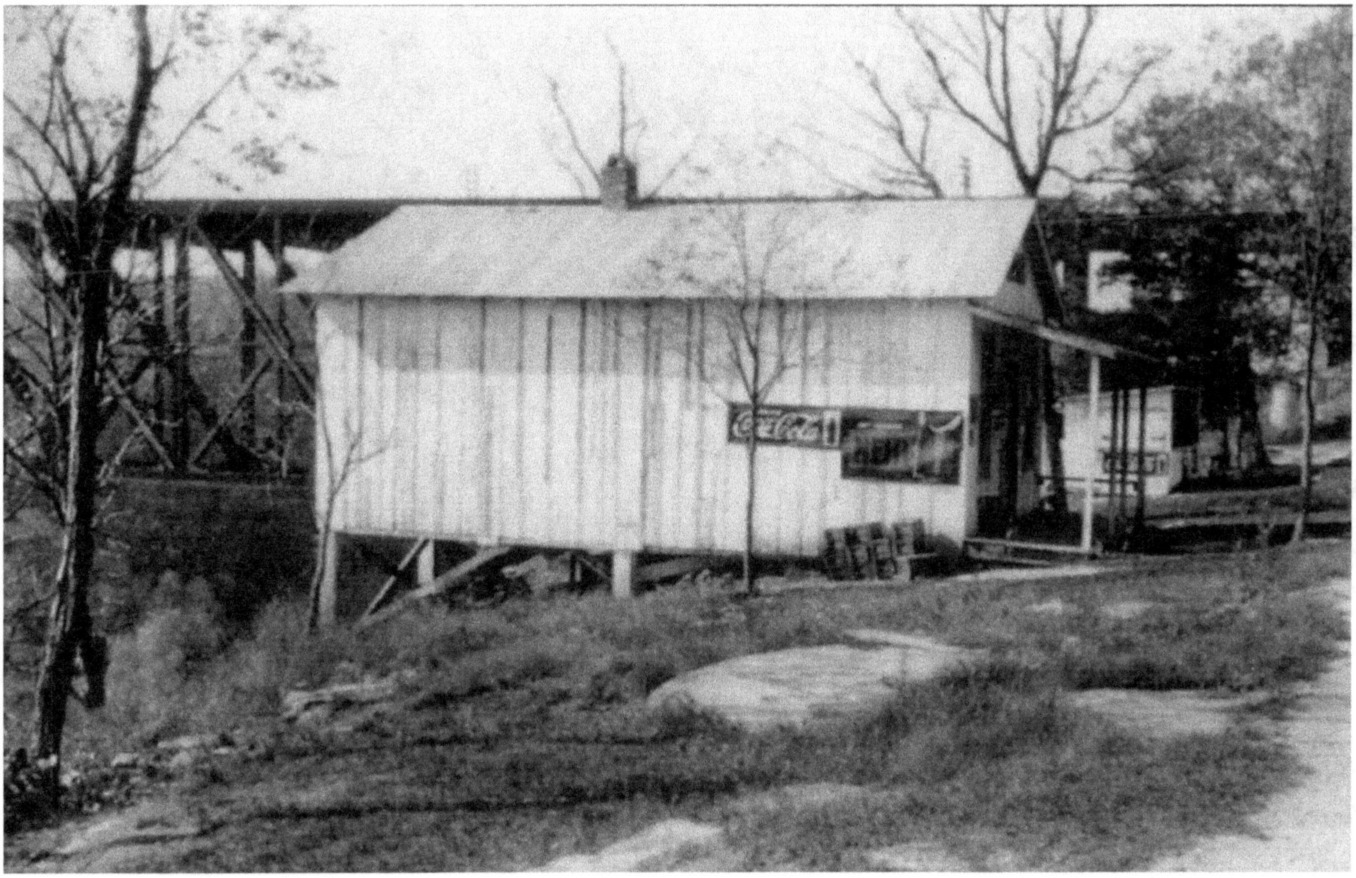

Located near the top of the stairs, this stand in High Bridge sold cold drinks to countless numbers of people on hot summer days, perhaps after they had just ascended the 271 stairs up the cliff from the riverbank below. Note the advertisements for Nehi Soda and Coca-Cola. The photograph was taken between 1924, the year Nehi Soda became available, and 1929.

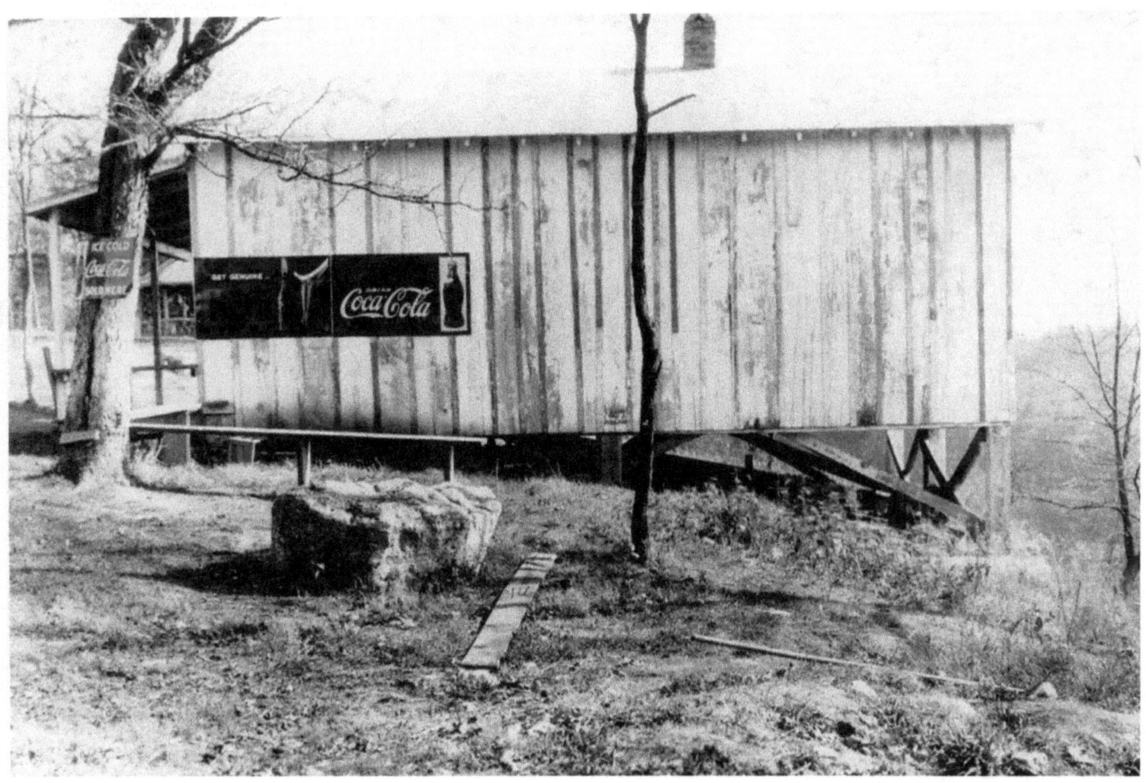

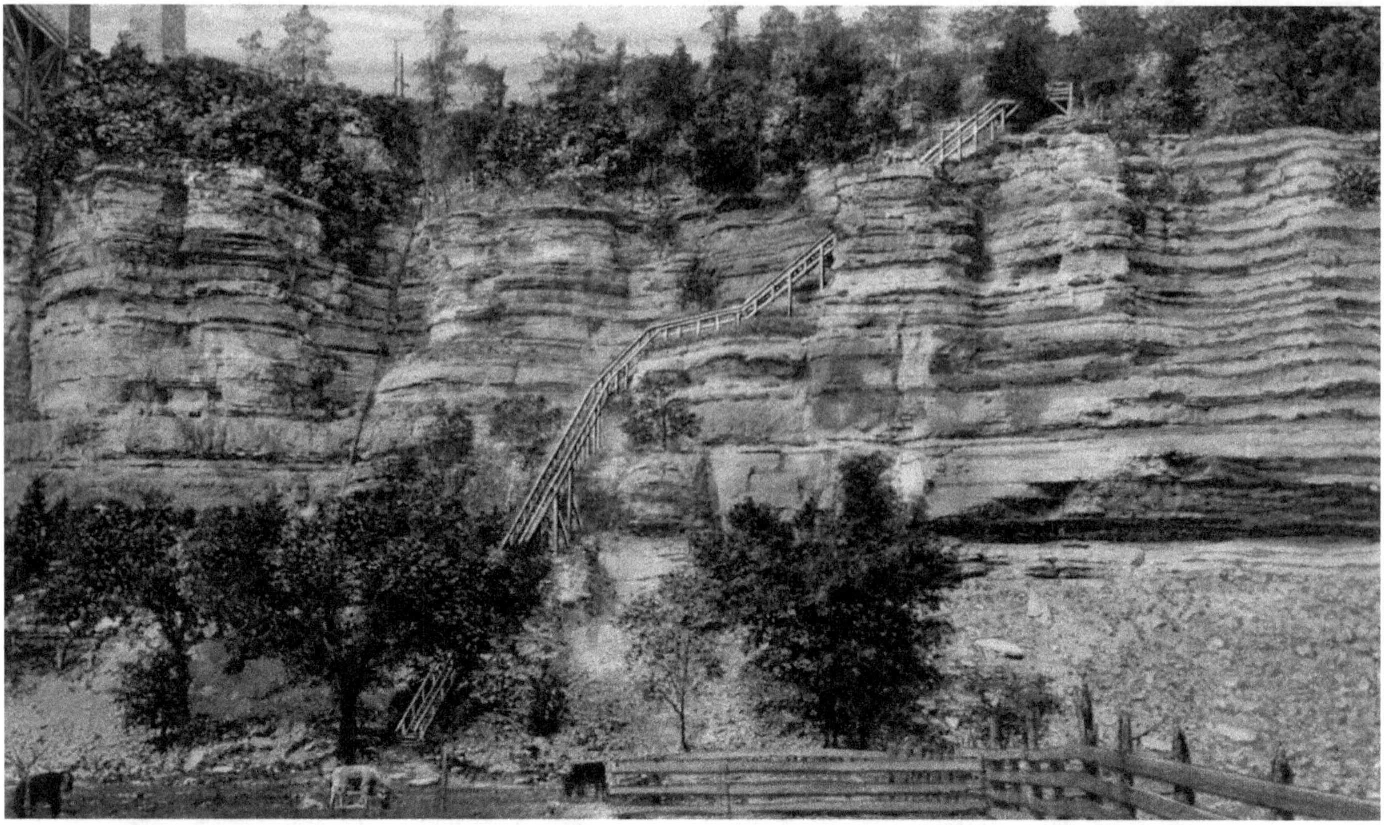

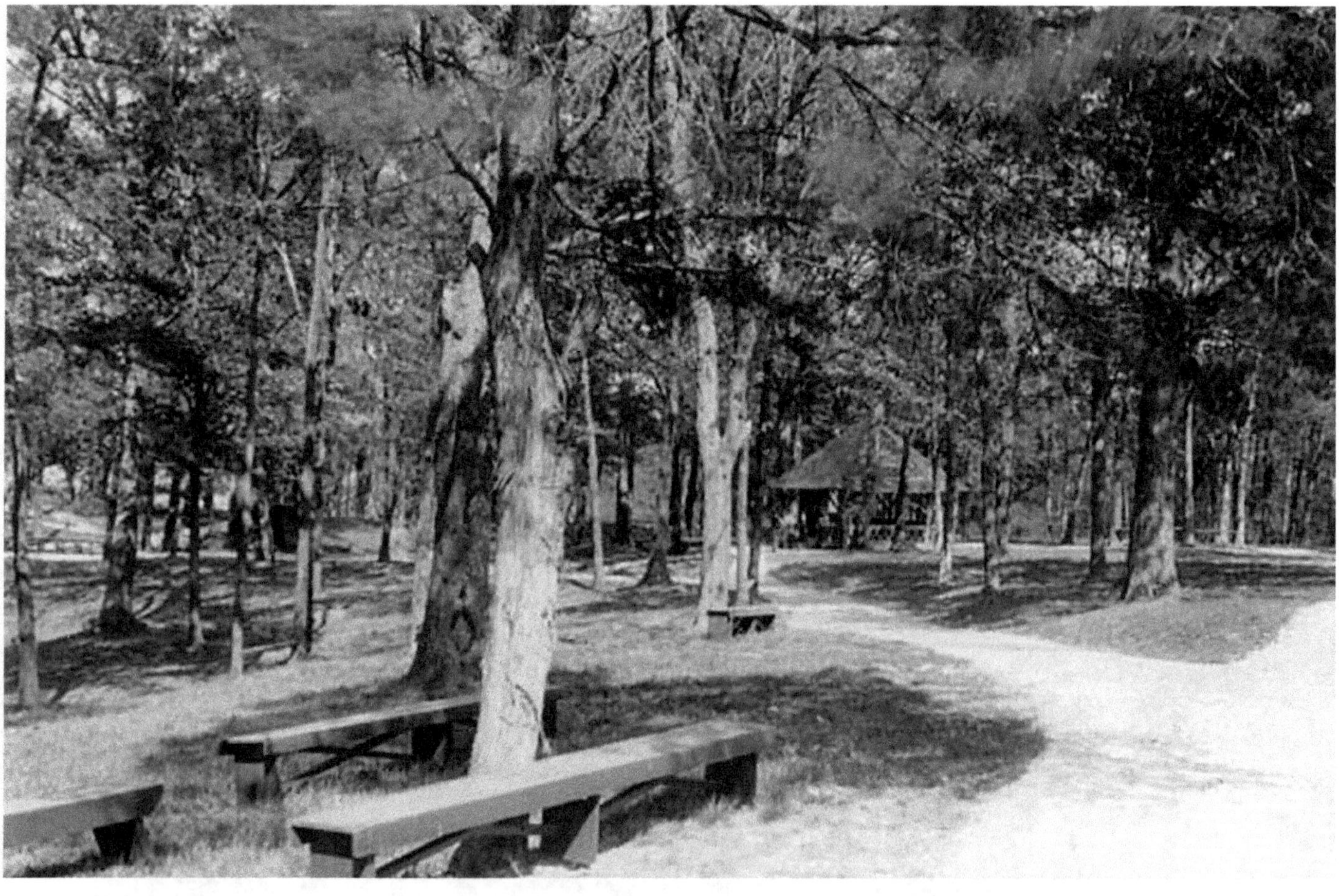

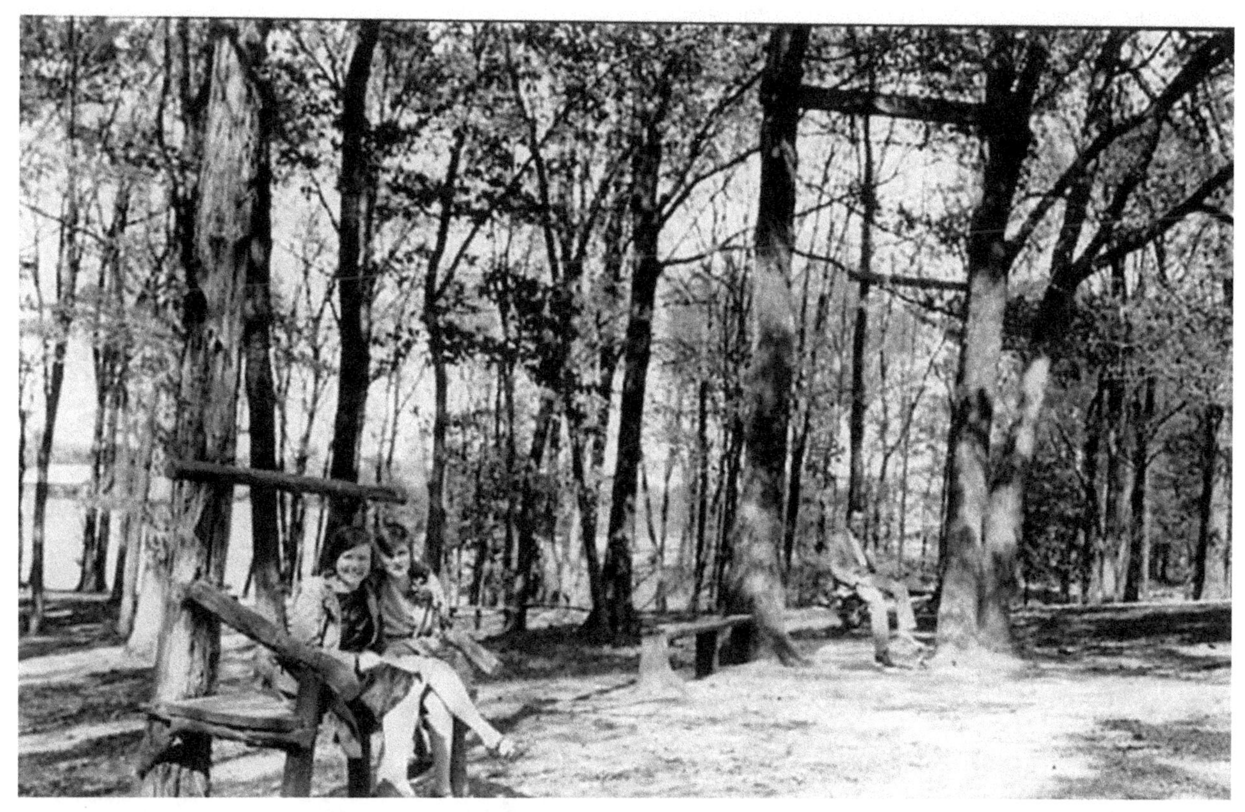

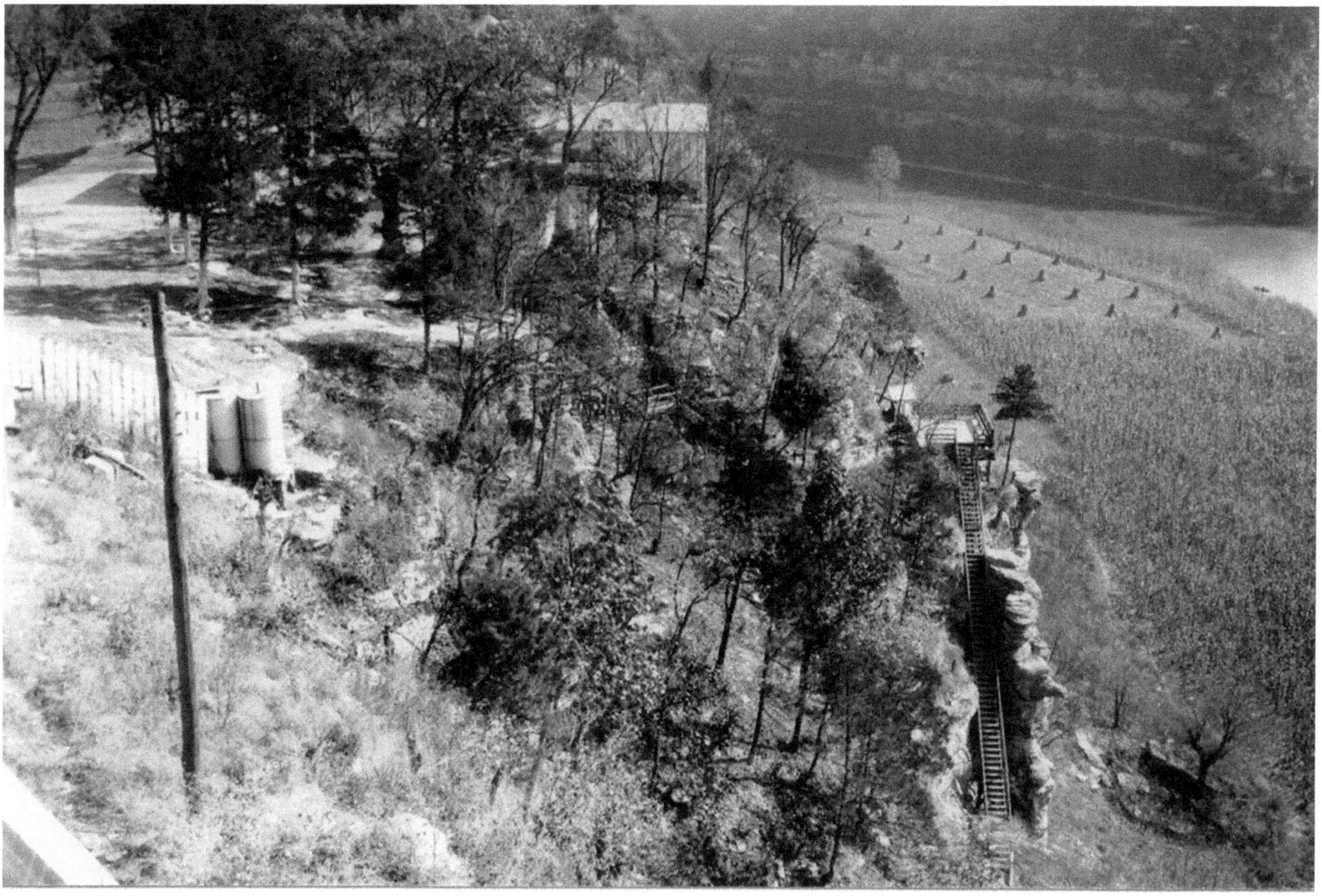

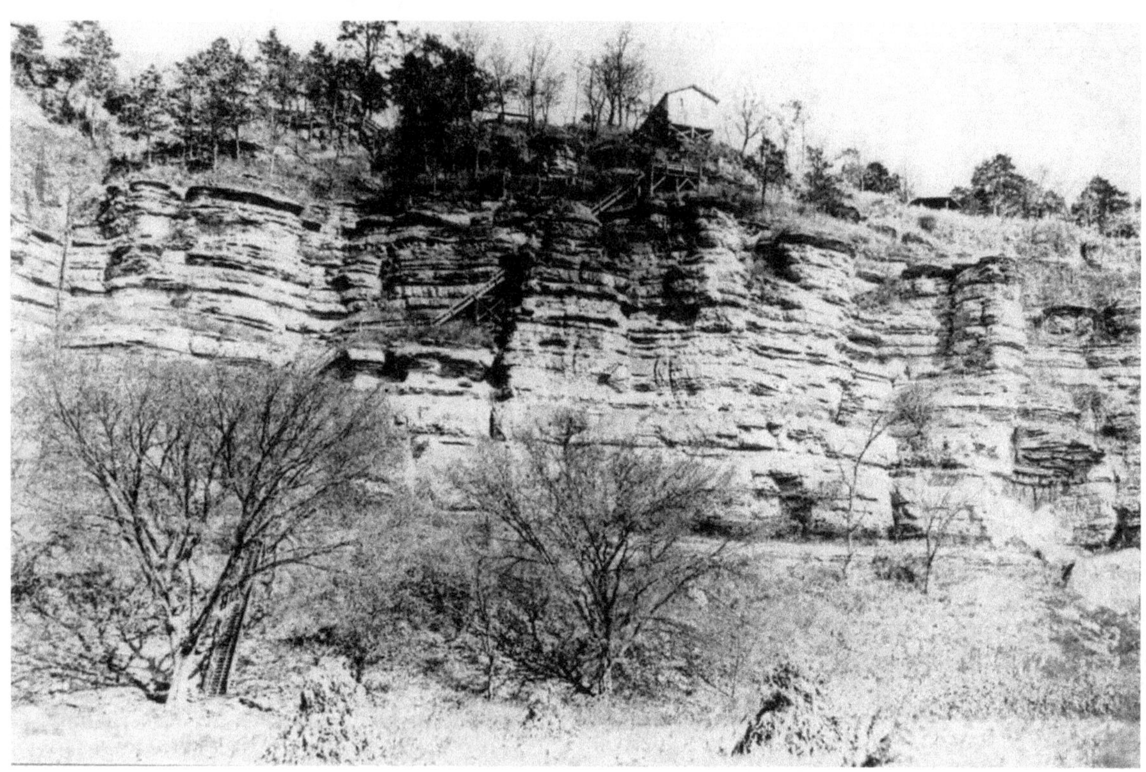

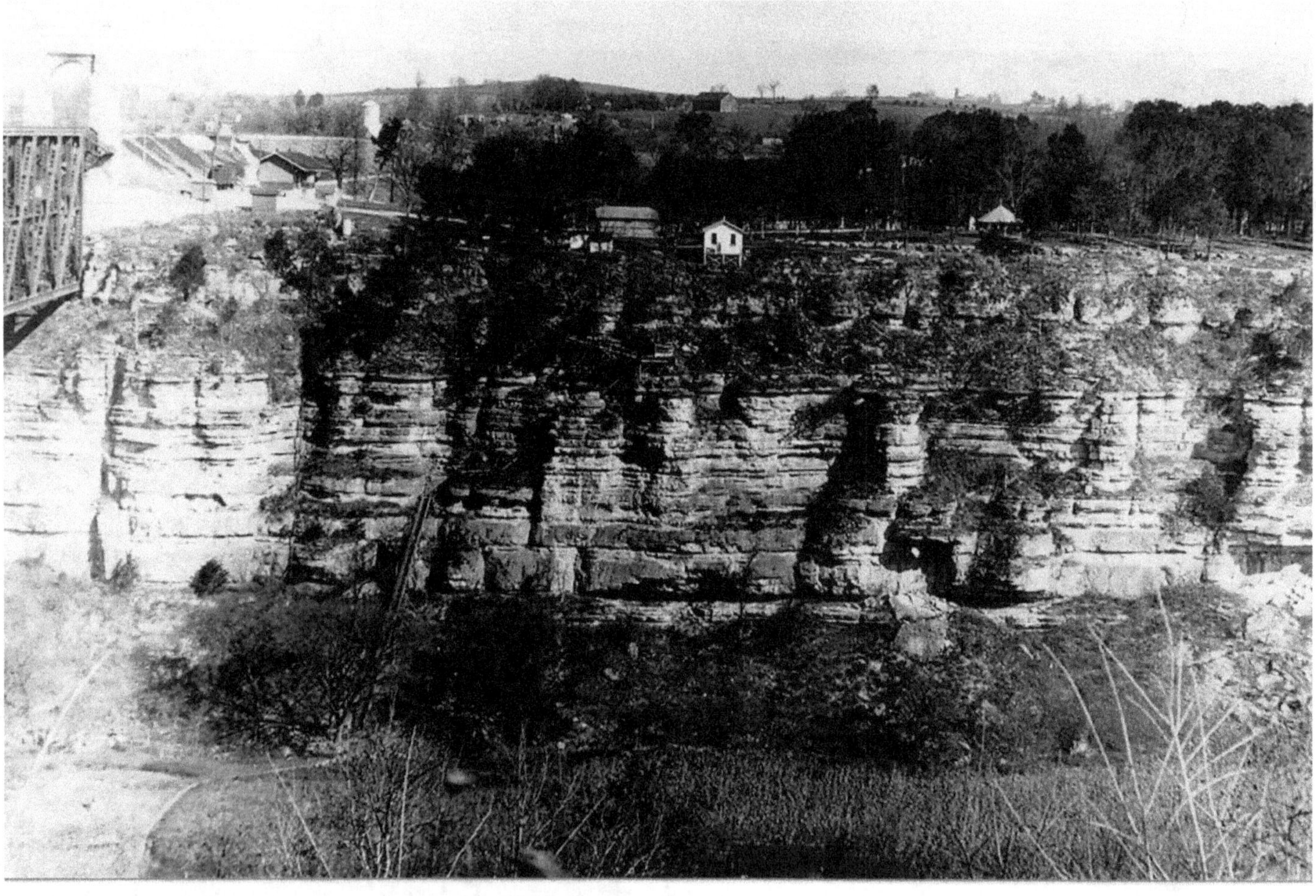

This is a unique vantage point looking from Mercer County across to the park in Jessamine County. On the far left of the photograph one can see the bridge and the depot in its new position. The building in the center of the photograph is the refreshment stand. One of the park's gazebos can be seen as well as the stairs down the cliff face.

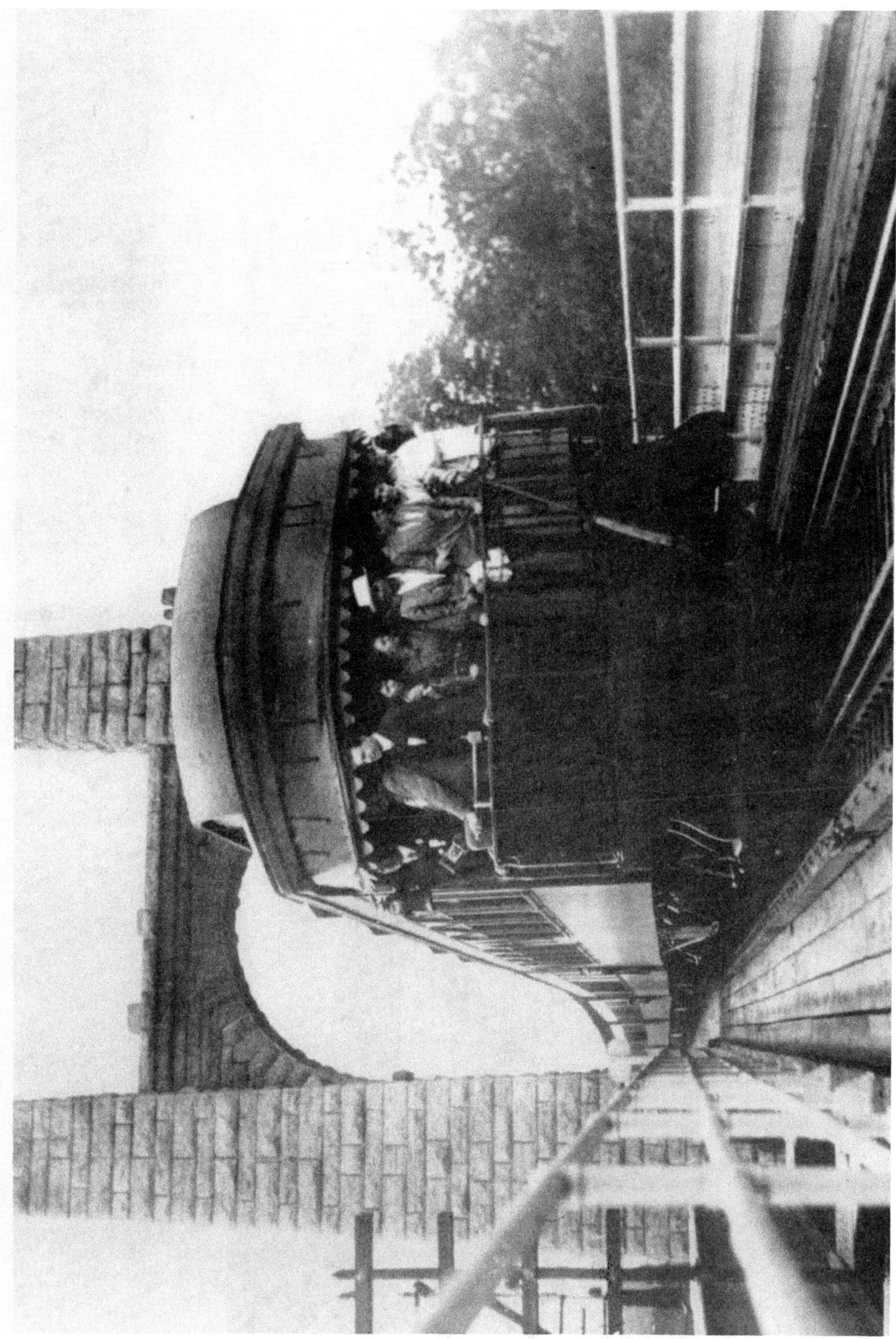

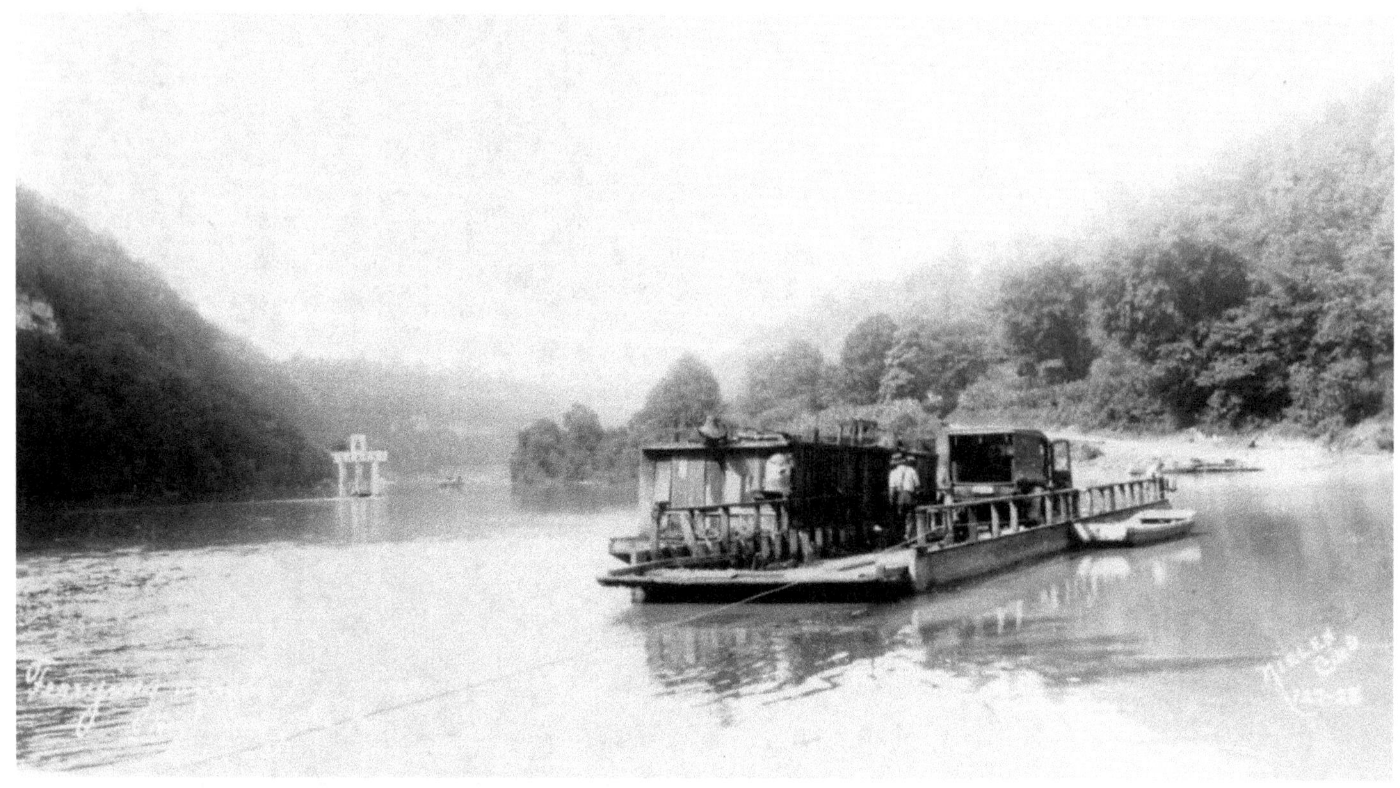

Seen in the distance in this view of the ferry operating at the Shaker Ferry Landing, Lock No. 7 was the only lock on the Kentucky River to have a self-contained hydroelectric power plant, which began commercial service in April 1928. The plant is still in operation, but is limited depending on the stage of the river.

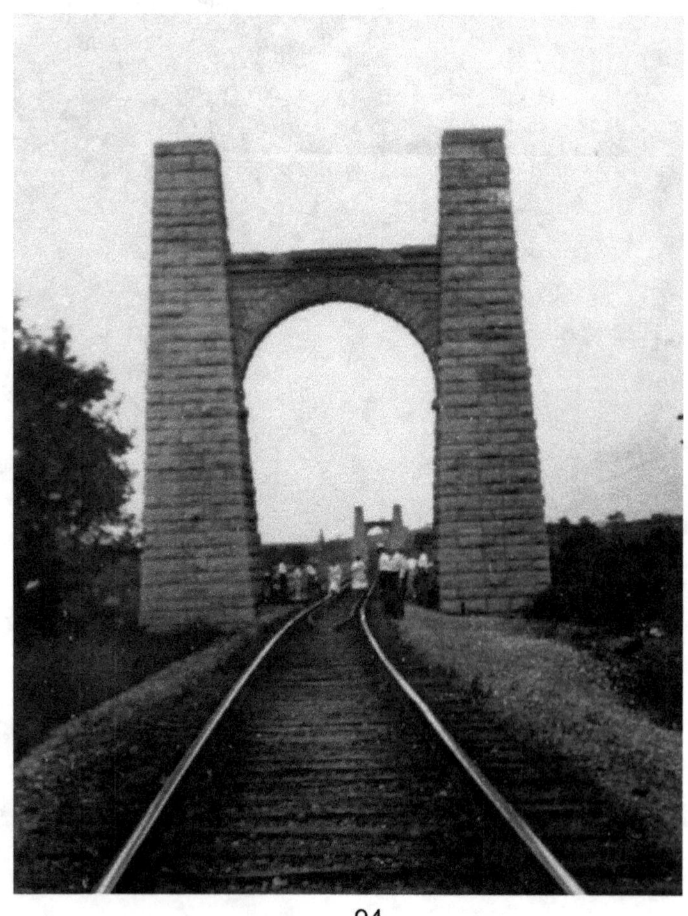

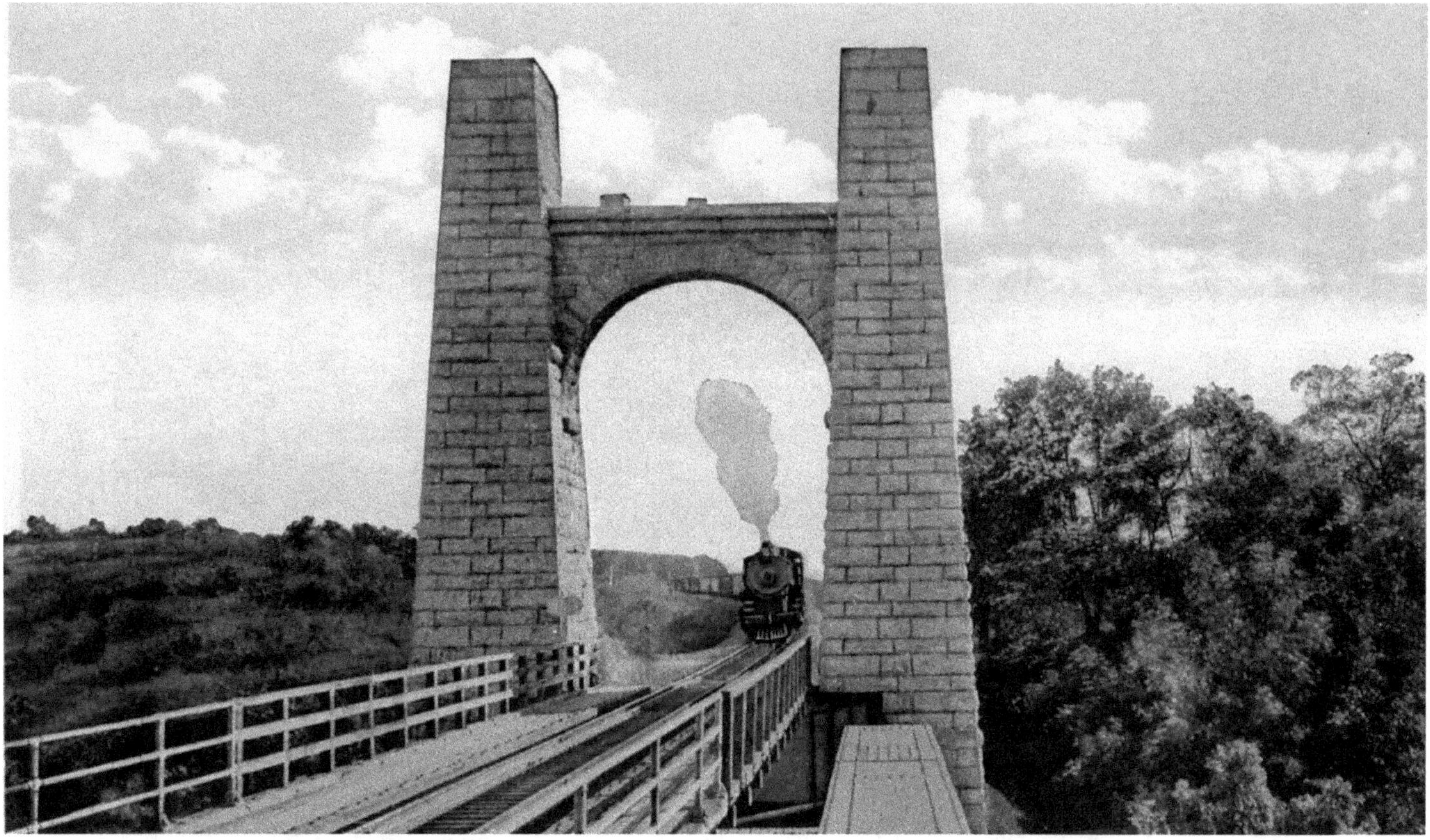

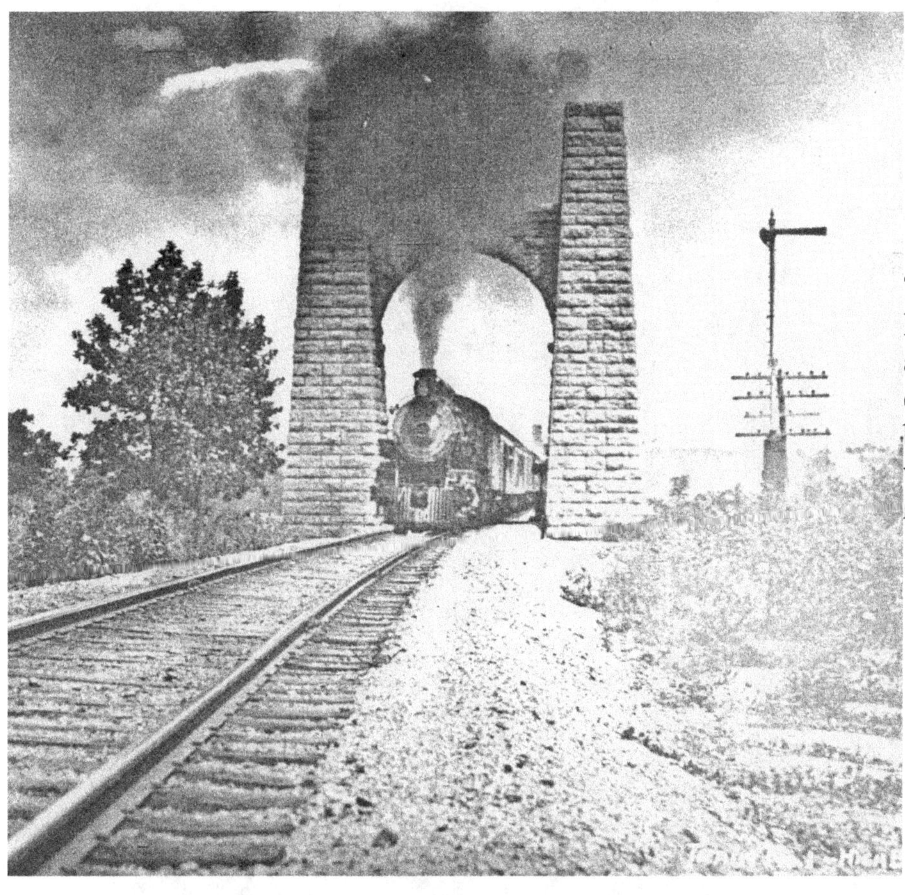

The raised level of the bridge gave Roebling's landmark towers a squatty appearance. The highest level of trains that crossed the bridge would pass about ten feet below the arch connecting the two towers, which made some passengers and onlookers nervous.

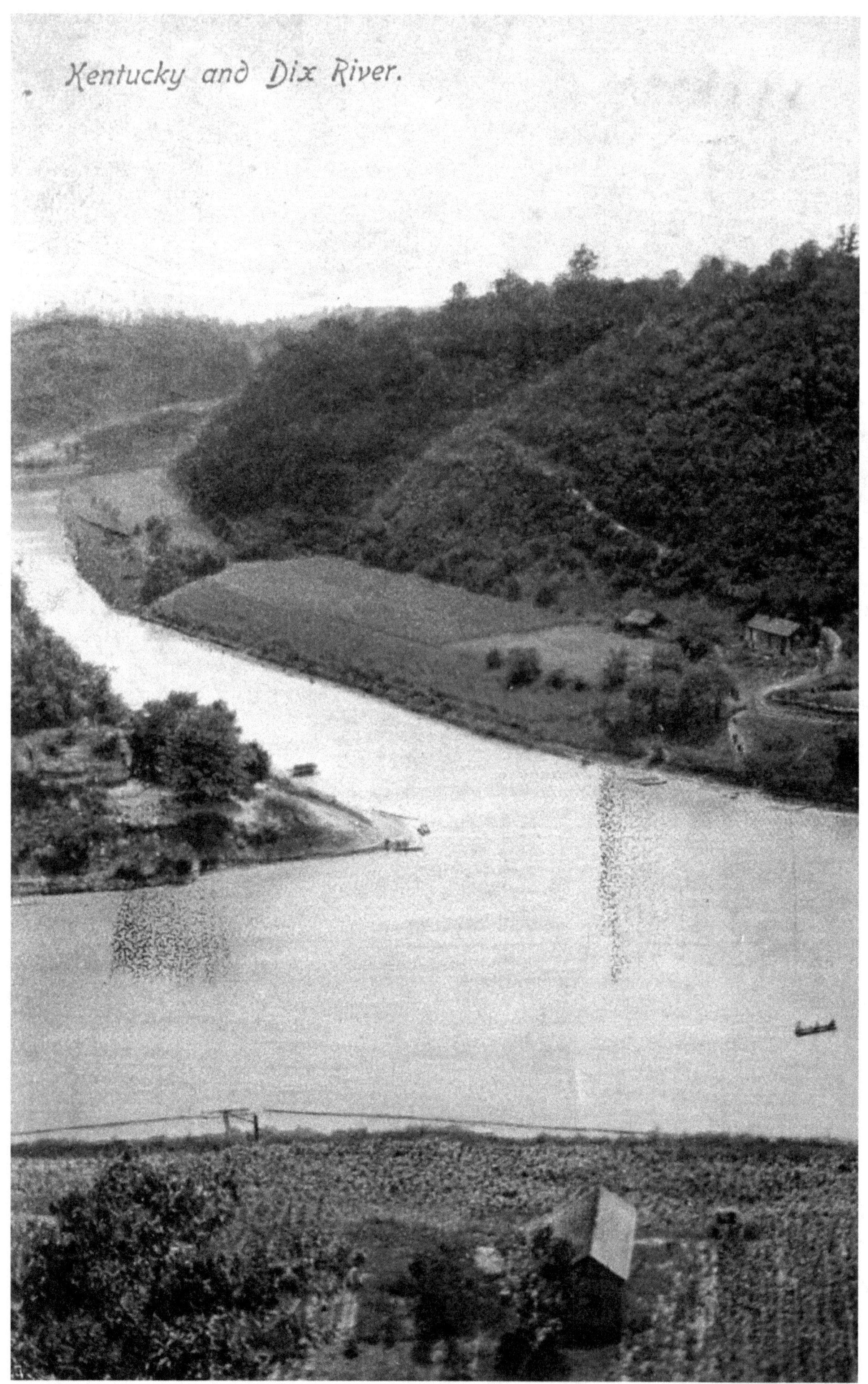
Kentucky and Dix River.

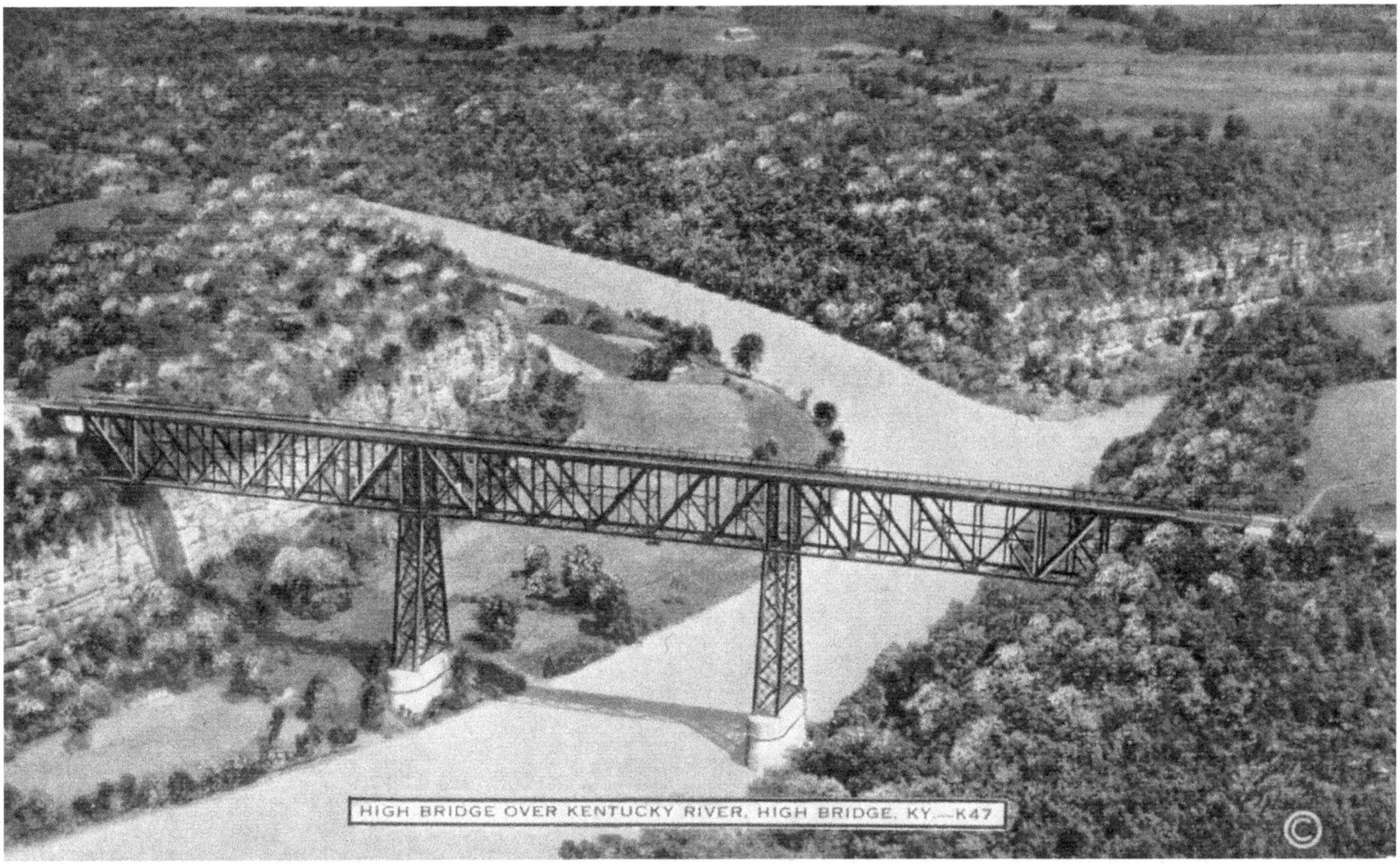

HIGH BRIDGE OVER KENTUCKY RIVER, HIGH BRIDGE, KY.—K47

The new bridge effortlessly carried traffic across the river, but once again the bridge needed modification. This time it was not for structural reasons, but to accommodate more frequent rail travel. In 1929 plans were made to lay two lanes of track across the bridge, one for northbound travel and one for southbound travel. But this modification doomed the landmark towers that Jonathan Roebling constructed in 1854, and the portion that towered over the tracks was removed.

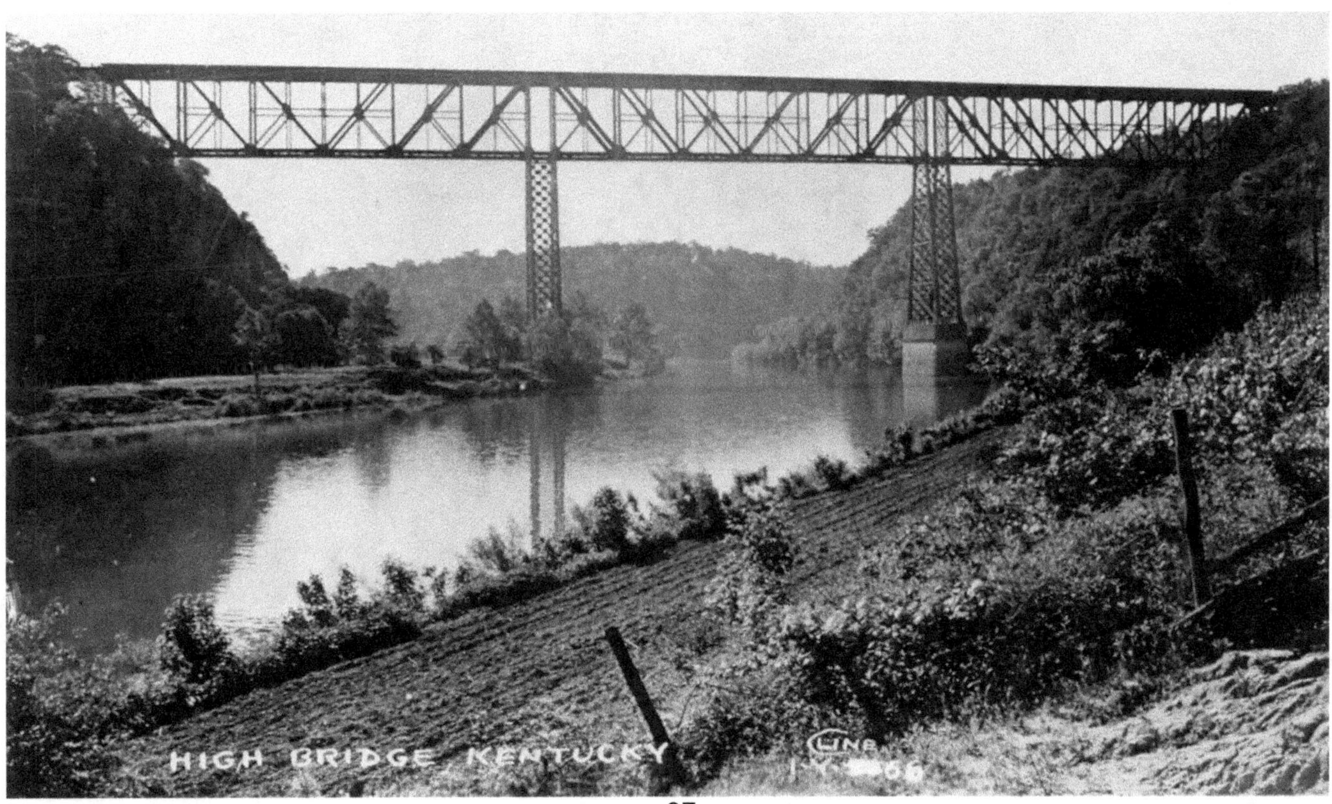

High Bridge's legendary stairway weathered the great flood of 1937 without sustaining major structural damage, only to be accidentally destroyed by railroad workers in 1948 who were burning brush along the riverbank. Due to the steady decline in visitors to the area, the stairs were never rebuilt.

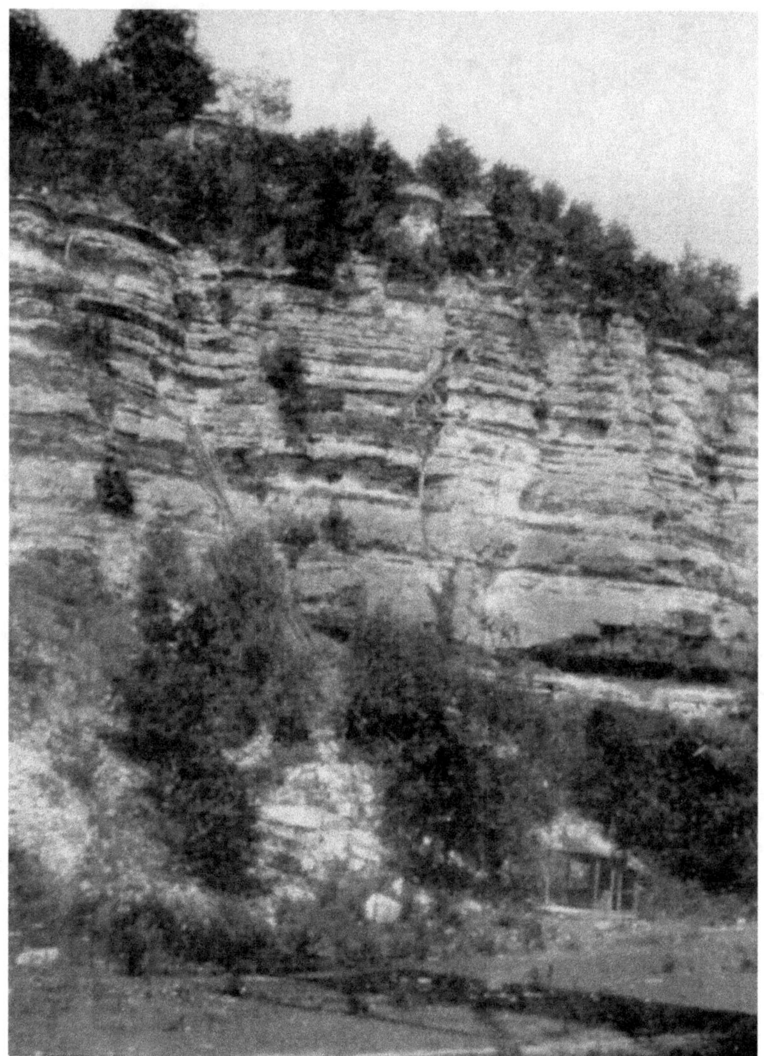

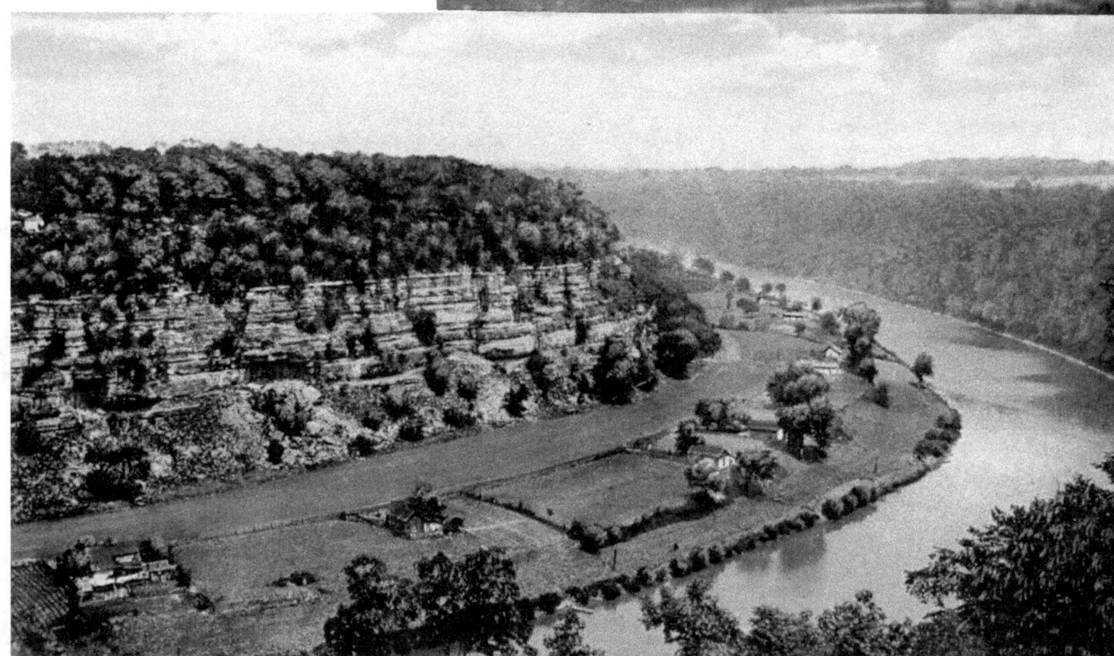

Seen here in the lower left, Ken Houp recalls many summer days spent at the home of his grandmother, Hadgie Mae Beckham Horn, in High Bridge while growing up. Her parents, Charles Beckham and Joanna Dearinger Beckham, had moved to High Bridge from the Cummins Ferry area in Mercer County many years before.

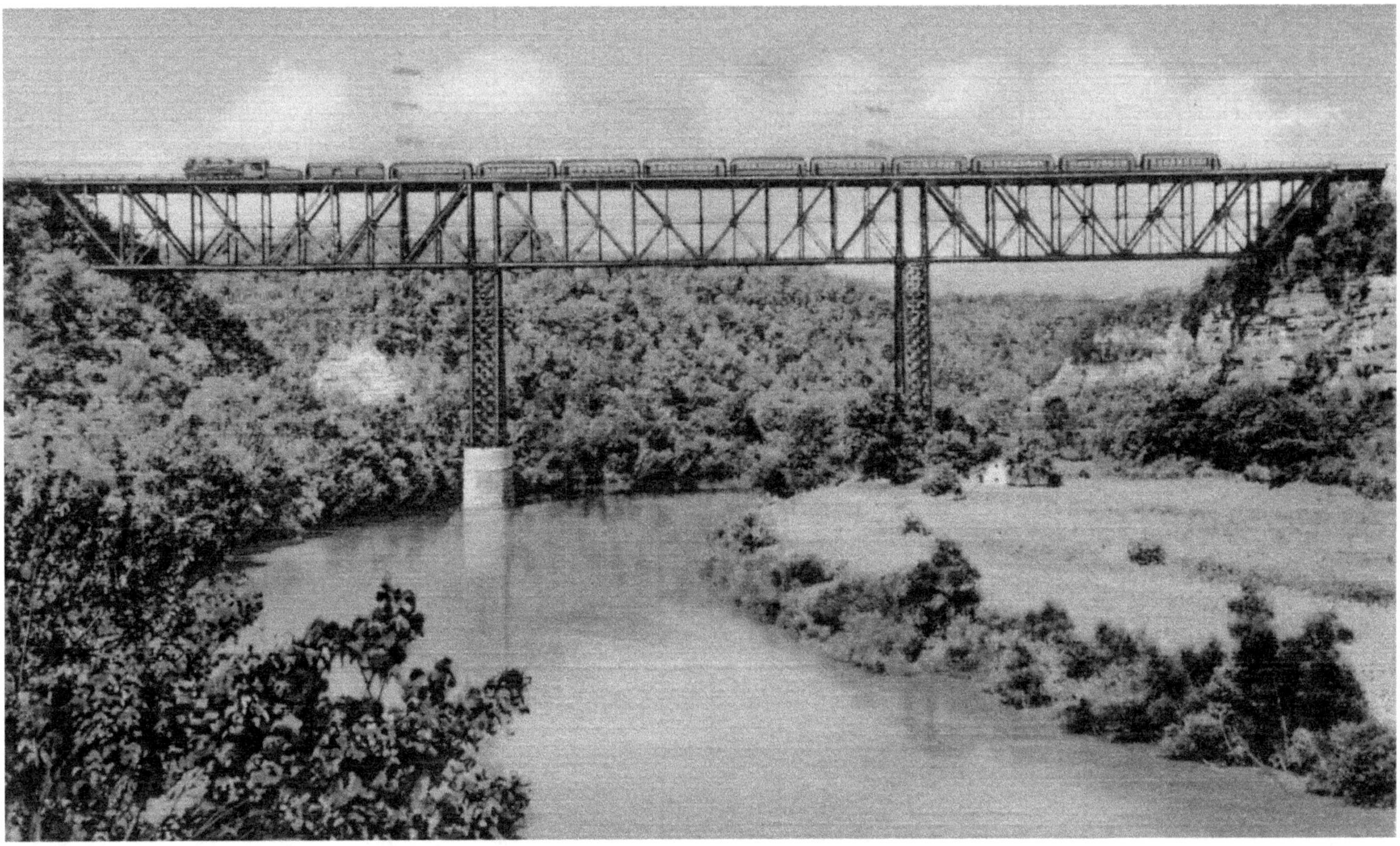

Ken Houp recalled a time when on the porch of his grandmother's home, there was a train crossing the bridge and blowing the tune "Shave and a Haircut" with its whistle. When he asked about it, she told him that it was her nephew Elmer "Bud" Horn. As a boy he had told her he wanted to drive trains across the bridge, and now that he was a train engineer, he would play the tune so she would know it was him crossing.

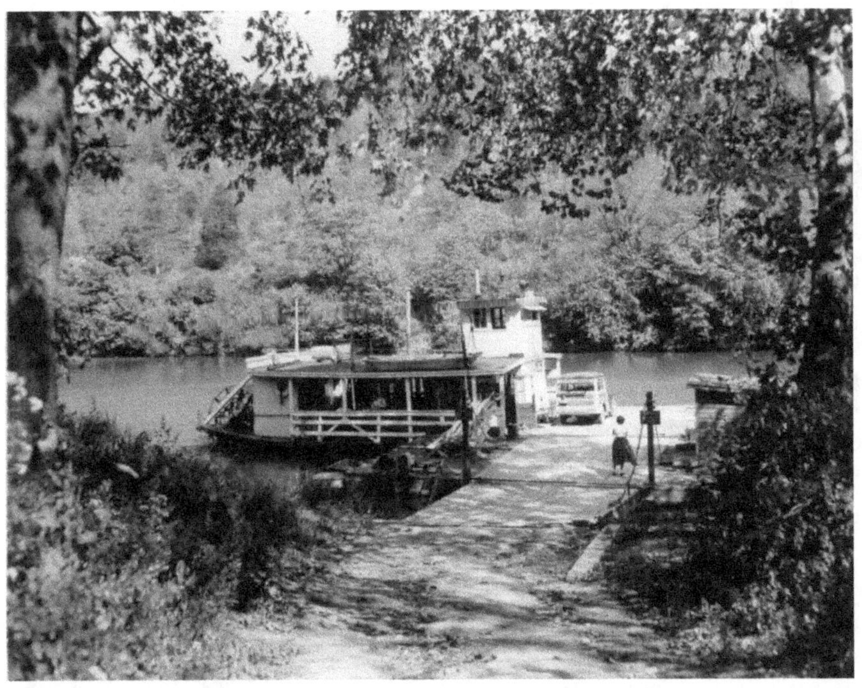

As horses and buggies gave way to automobiles, so the Kentucky back roads where ferries once operated gave way to more modern roads with bridges over waterways. The ferry that operated out of Shaker Ferry Landing was one such victim. This photograph was taken near the end of its operation on the river.

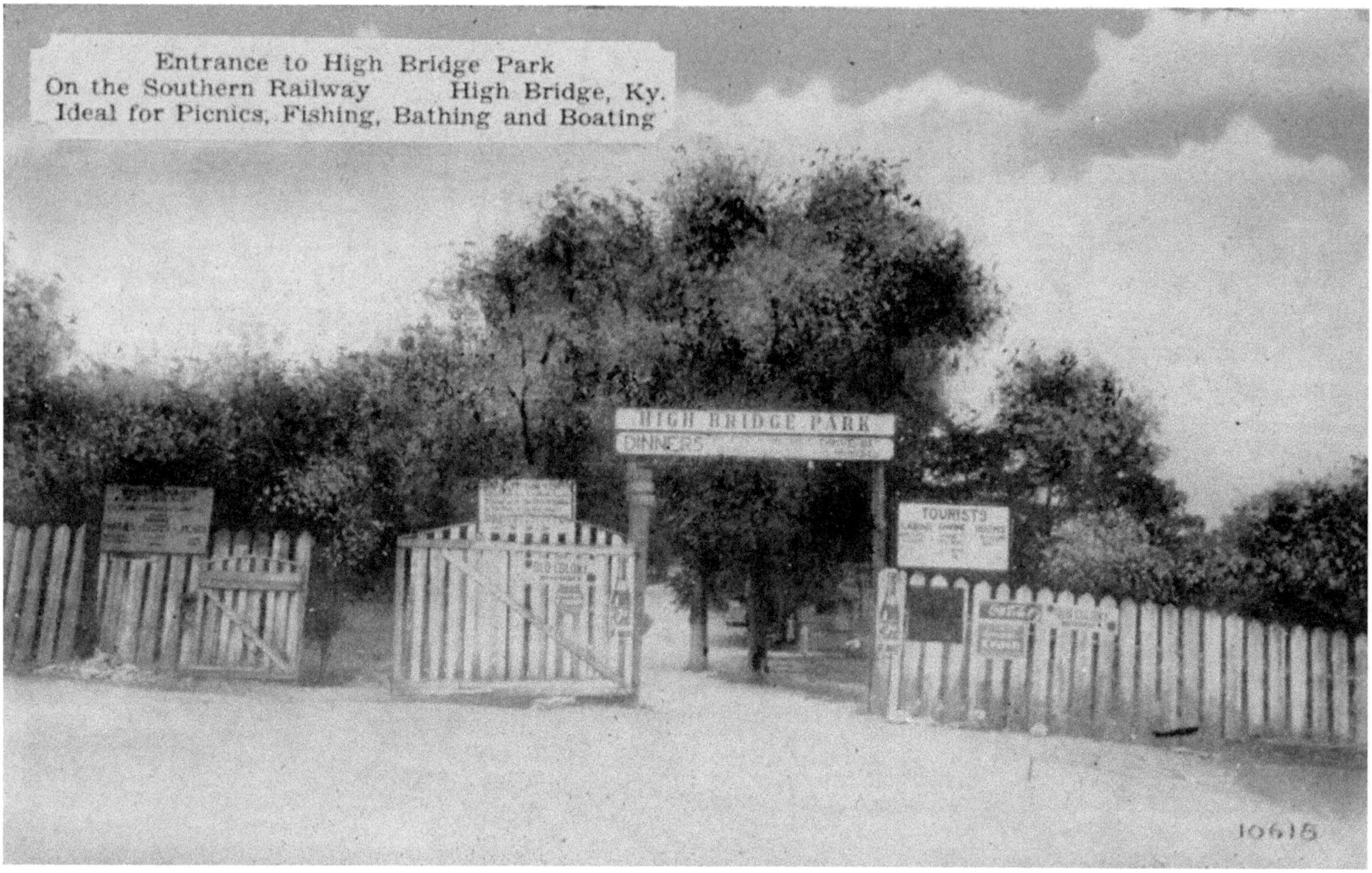

Entrance to High Bridge Park
On the Southern Railway High Bridge, Ky.
Ideal for Picnics, Fishing, Bathing and Boating

10615

The gates of High Bridge Park greeted many sightseers, young and old, for generations. High Bridge Park was once a place for innocence. Over time and due to social changes, the clientele that frequented the park became rougher and families quit frequenting the park. By the 1970's, the park was totally abandoned.

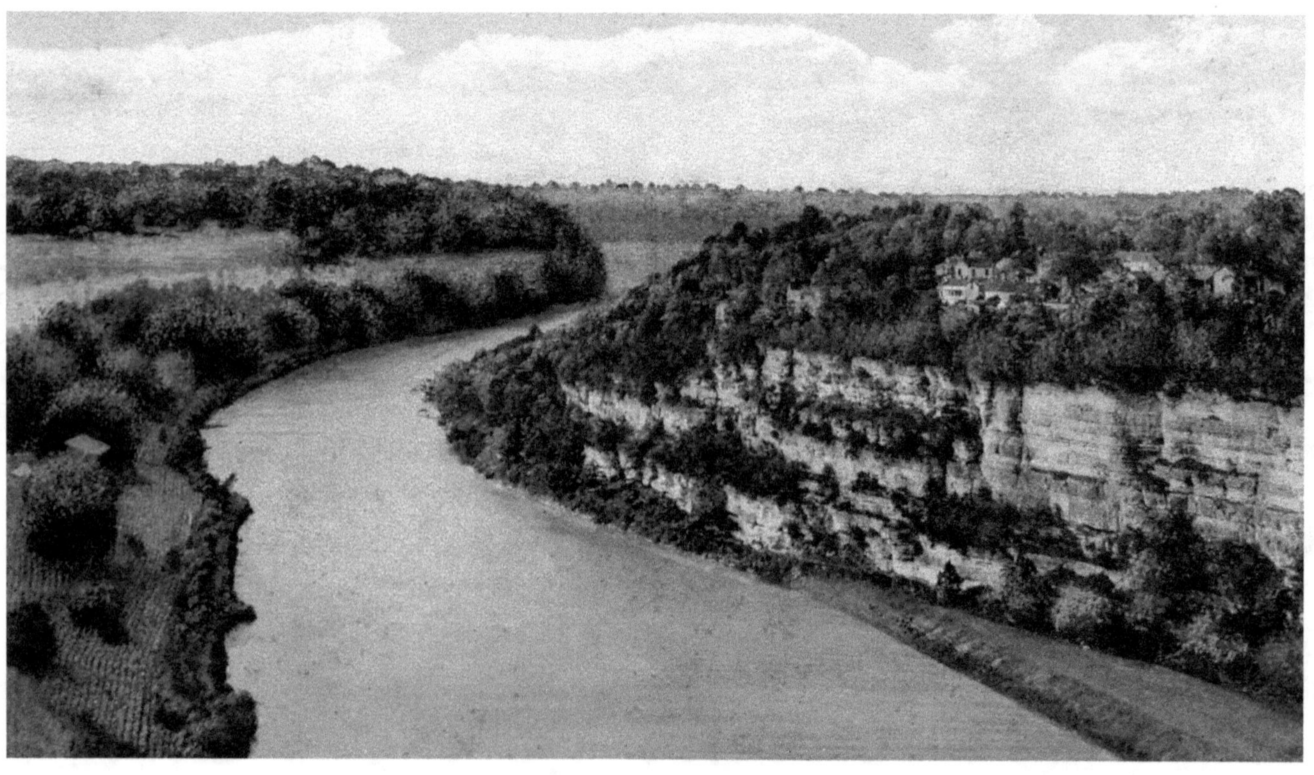

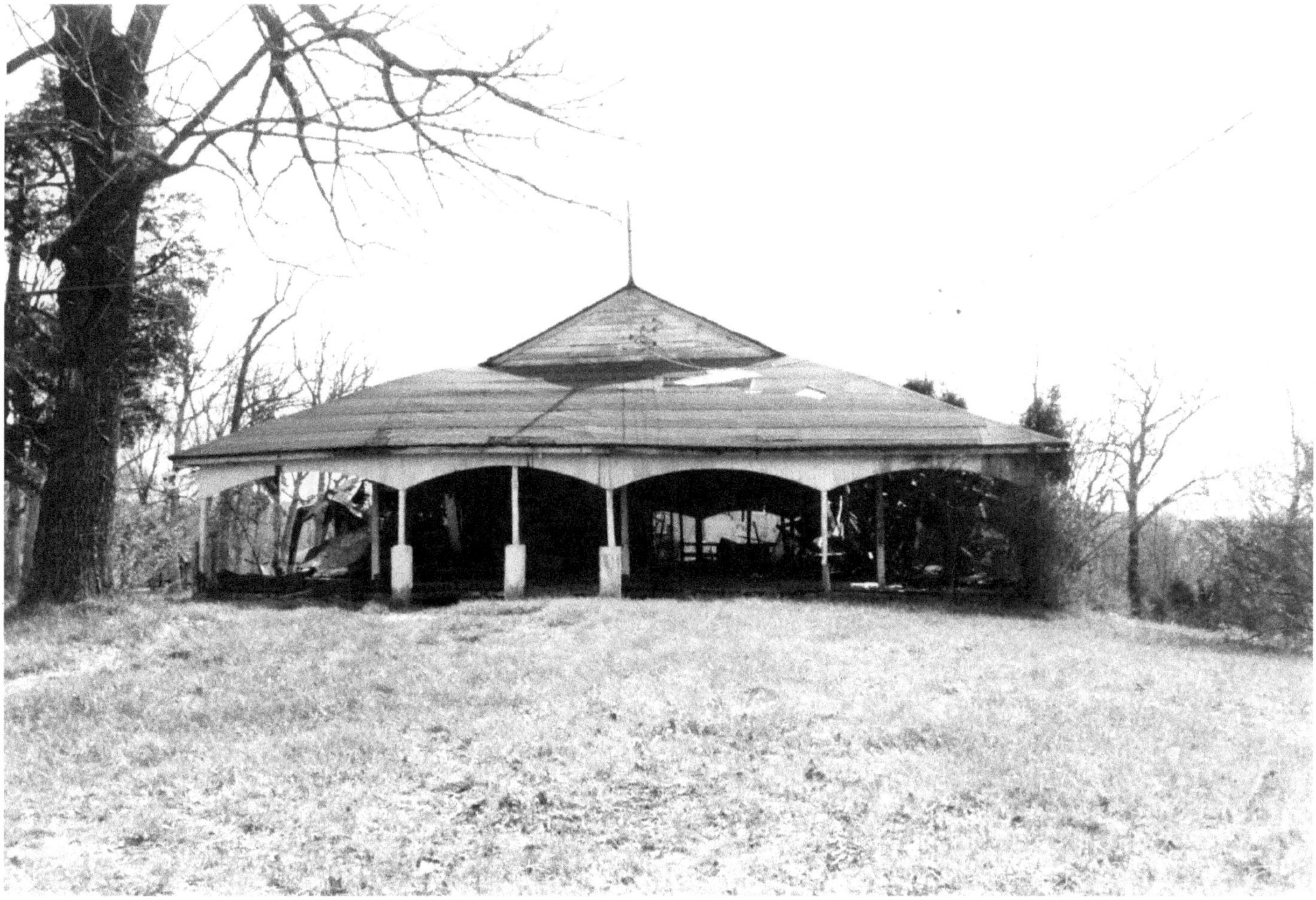

When High Bridge Park was officially closed in the 1970's, the structures in the park were left to the elements. Due to time, lack of maintenance, and vandalism, the old dance pavilion collapsed, perhaps longing for the days when dancing feet glided across its wooden floor and laughter echoed in its rafters.

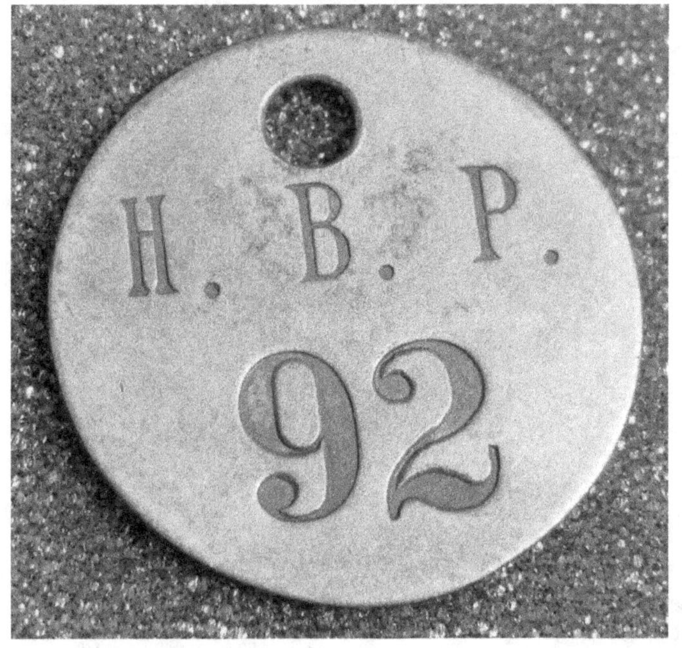

This brass coat check tag is a reminder that High Bridge Park used to host large gatherings and events of its day. Whoever had tag 92 must have retrieved their coat and decided to pocket the tag as a memento of their time in High Bridge Park, or accidentally left their coat behind!

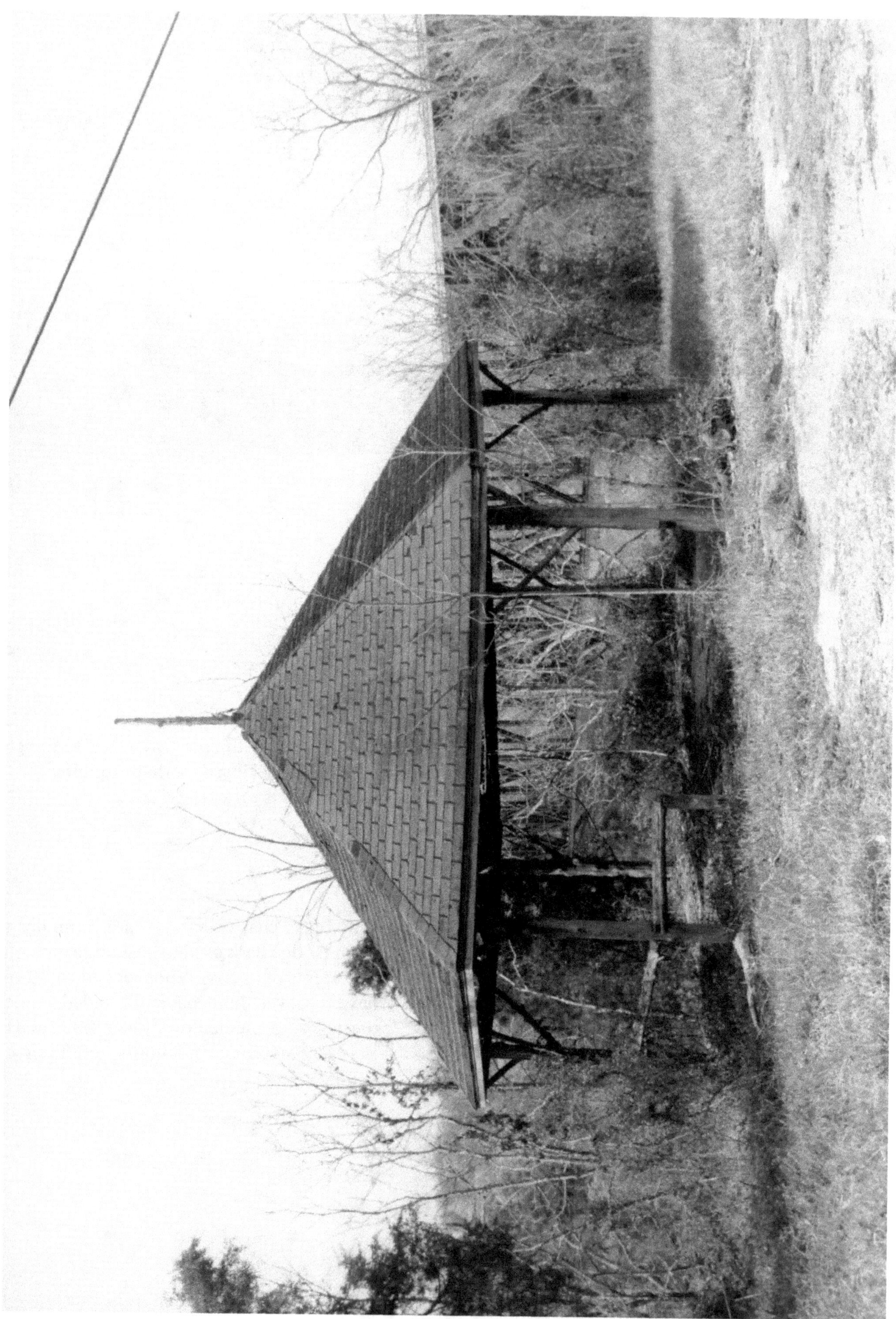

High Bridge Today

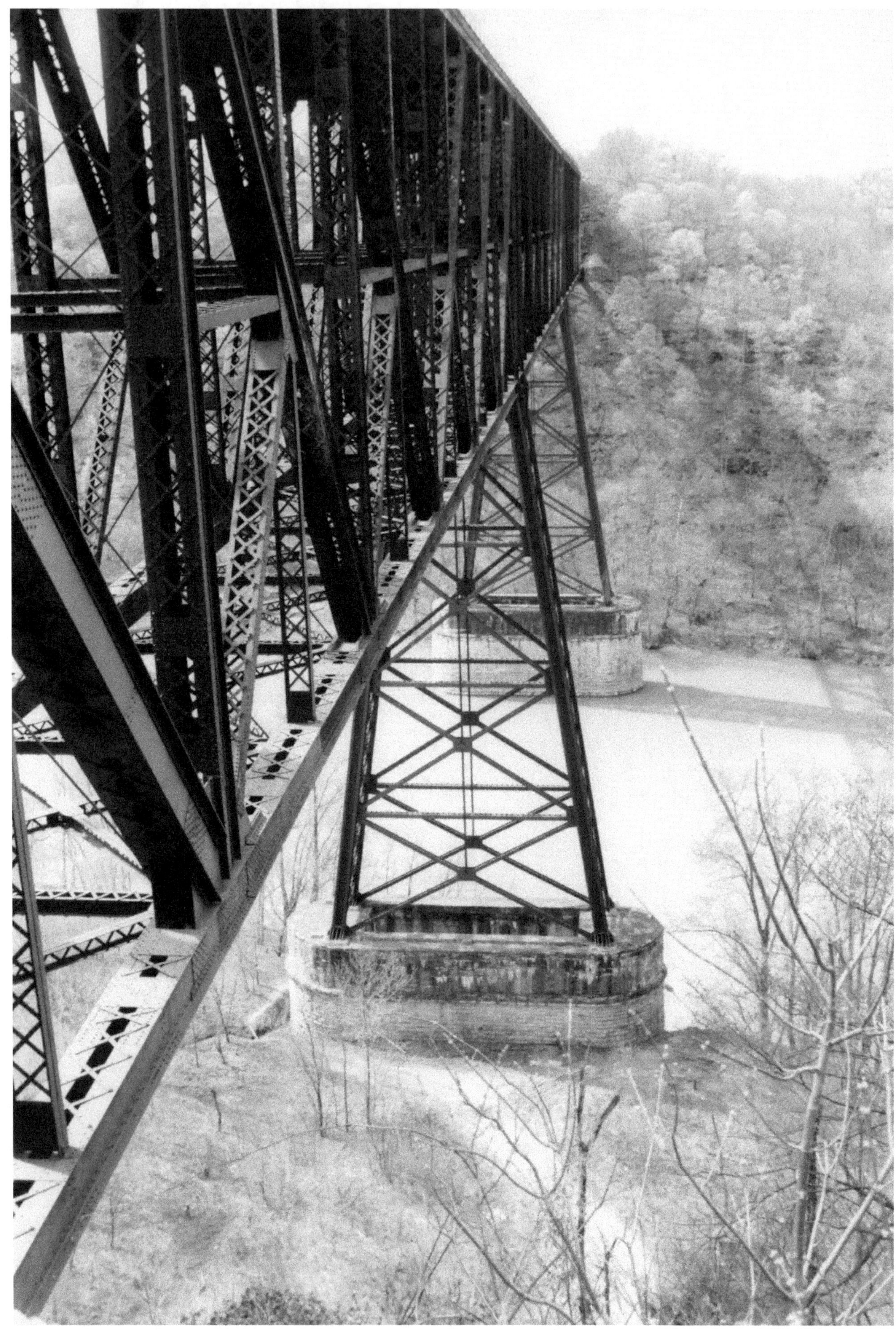

At its peak in the 1920's up to twelve passenger trains per day visited the bridge and park however, today none visit High Bridge. Gone are the days of the steamboats that brought passengers from afar to see the steel marvel. Due to the closure of the locks and dams along the Kentucky River by the U.S. Army Corps of Engineers, the river itself is no longer completely navigable. Despite these facts the Gustav Lindenthal's High Bridge has withstood the test of time and continues to carry the rails for years to come. This relic from our past still easily accommodates the heaviest of trains, with one crossing about every thirty minutes during peak hours, sometimes carrying a northbound and southbound train simultaneously. In fact, this bridge lies on the busiest north-south route on the Norfolk-Southern Railroad line in the nation.

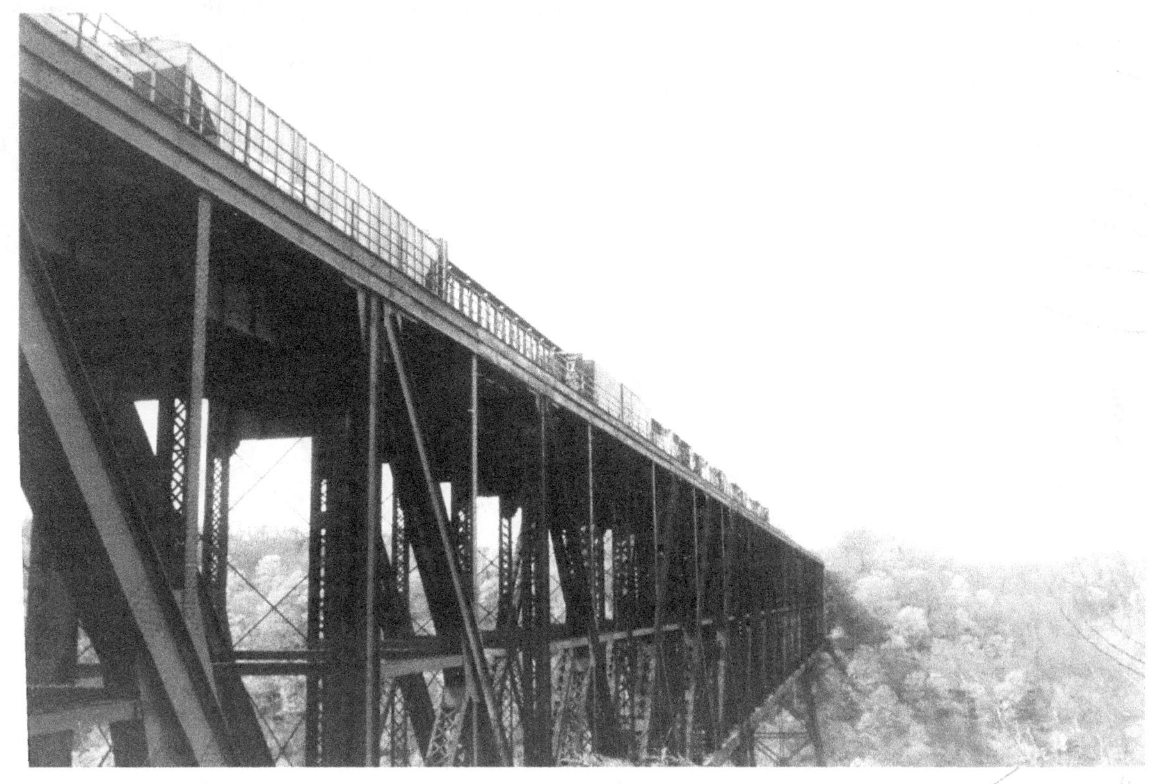

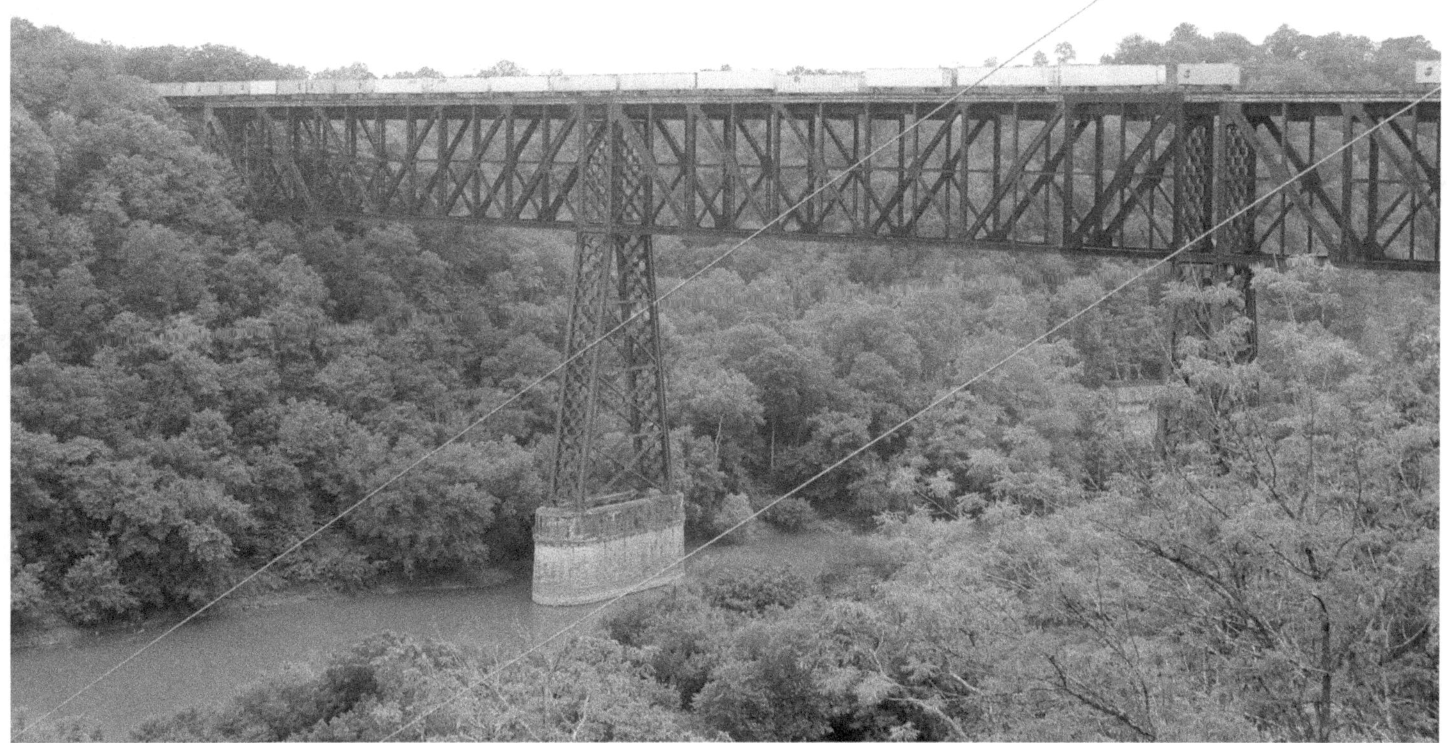

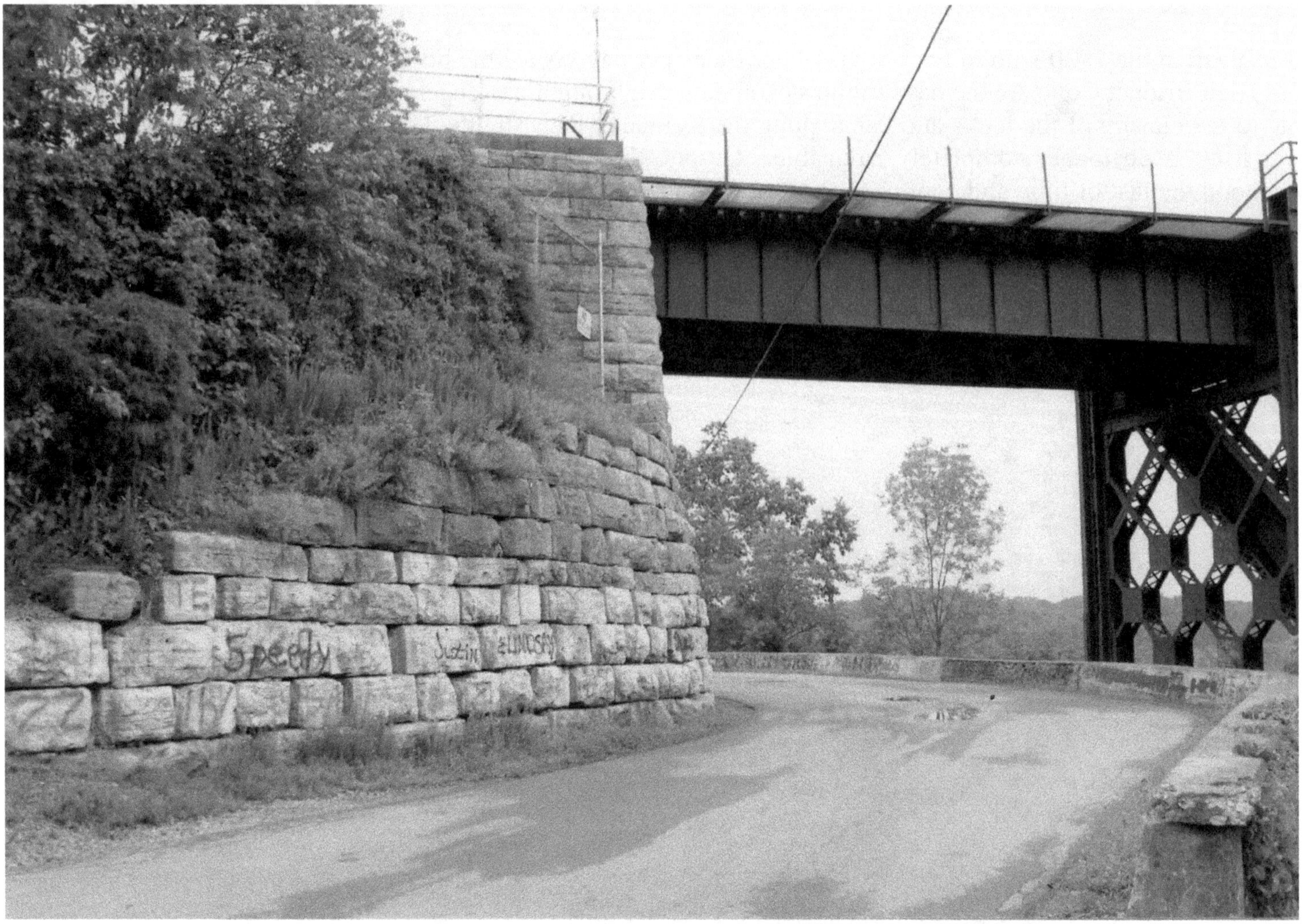

While it was widely reported that Roebling's stone towers were totally removed from High Bridge in 1929, the biggest portion, if not all of the towers, still remains in its original position. When the double track was laid, the stones above the bridge deck were relocated in a graduating stack to act as a barrier between the ballast and the roadway.

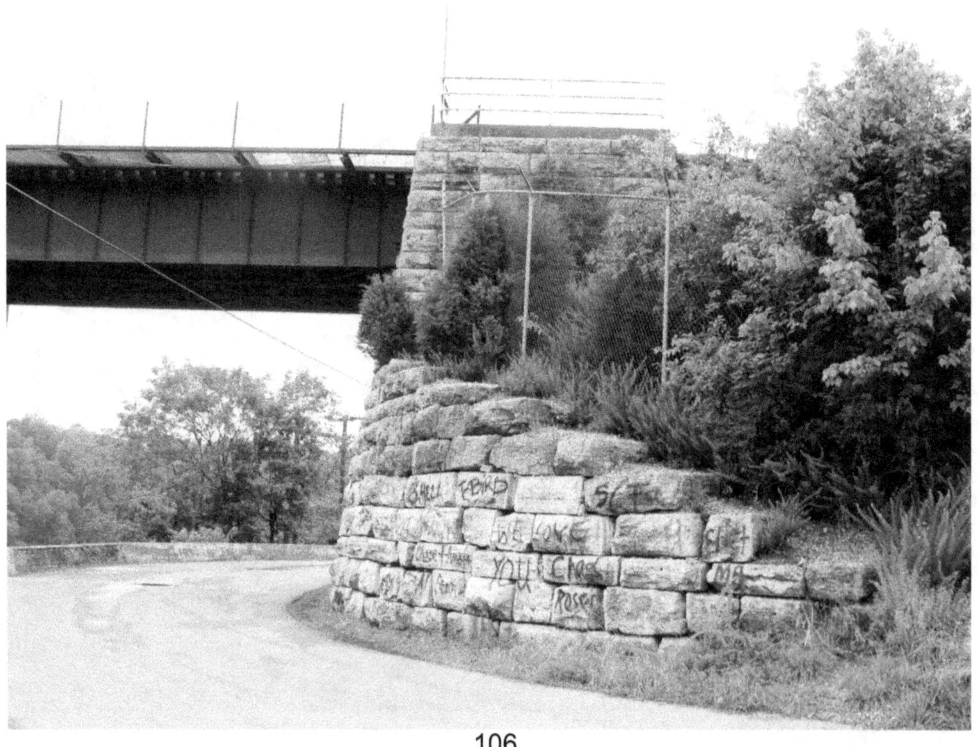

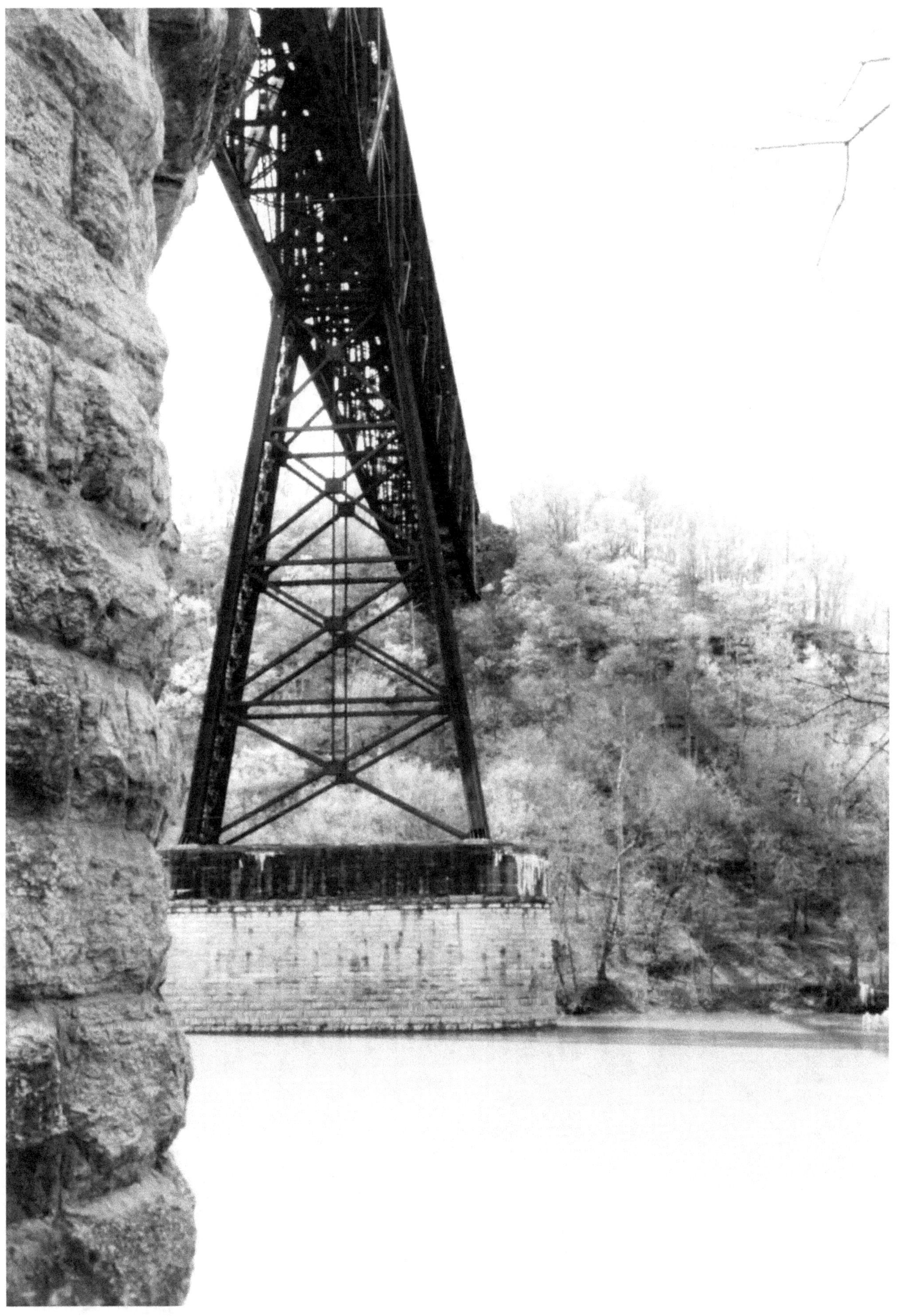

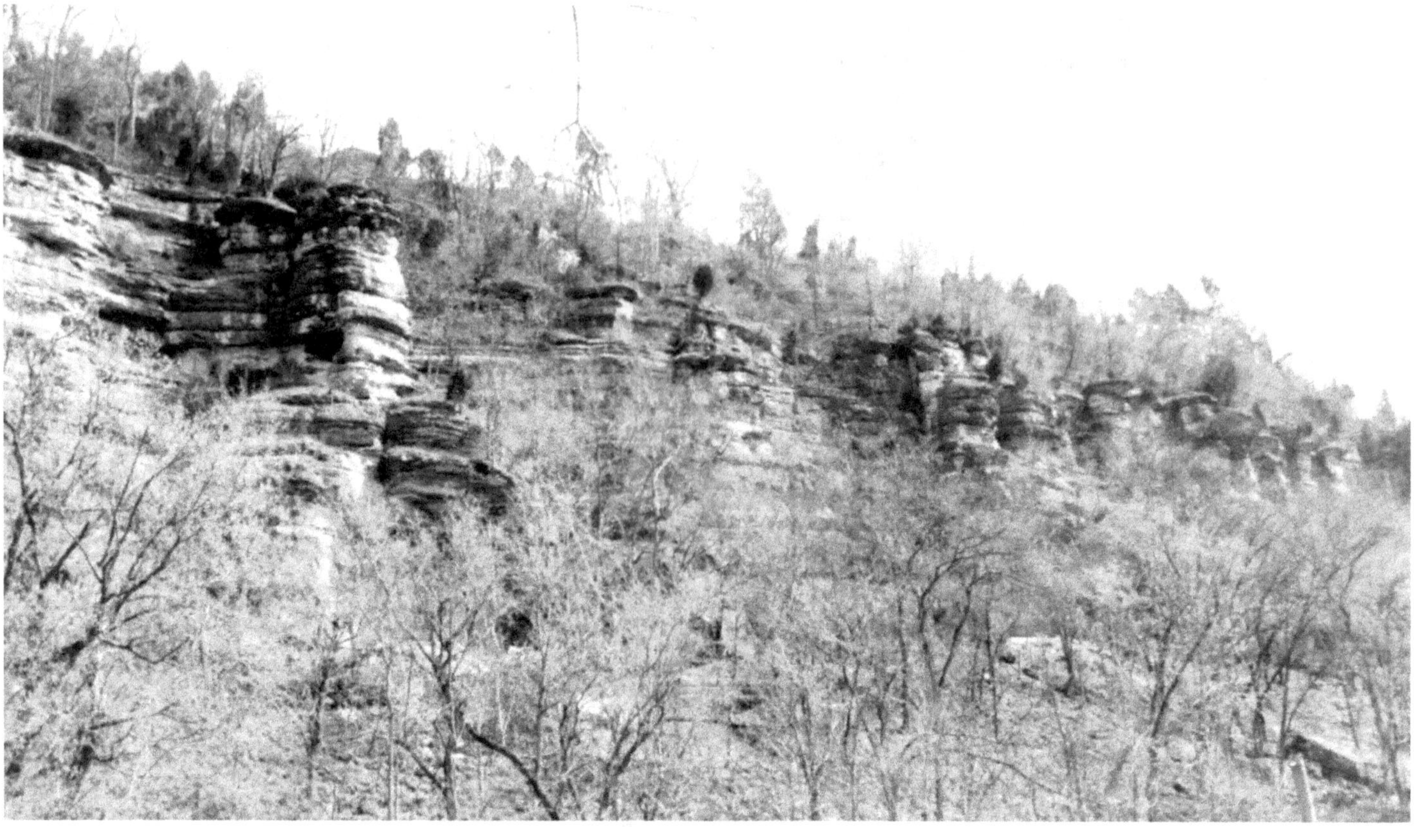

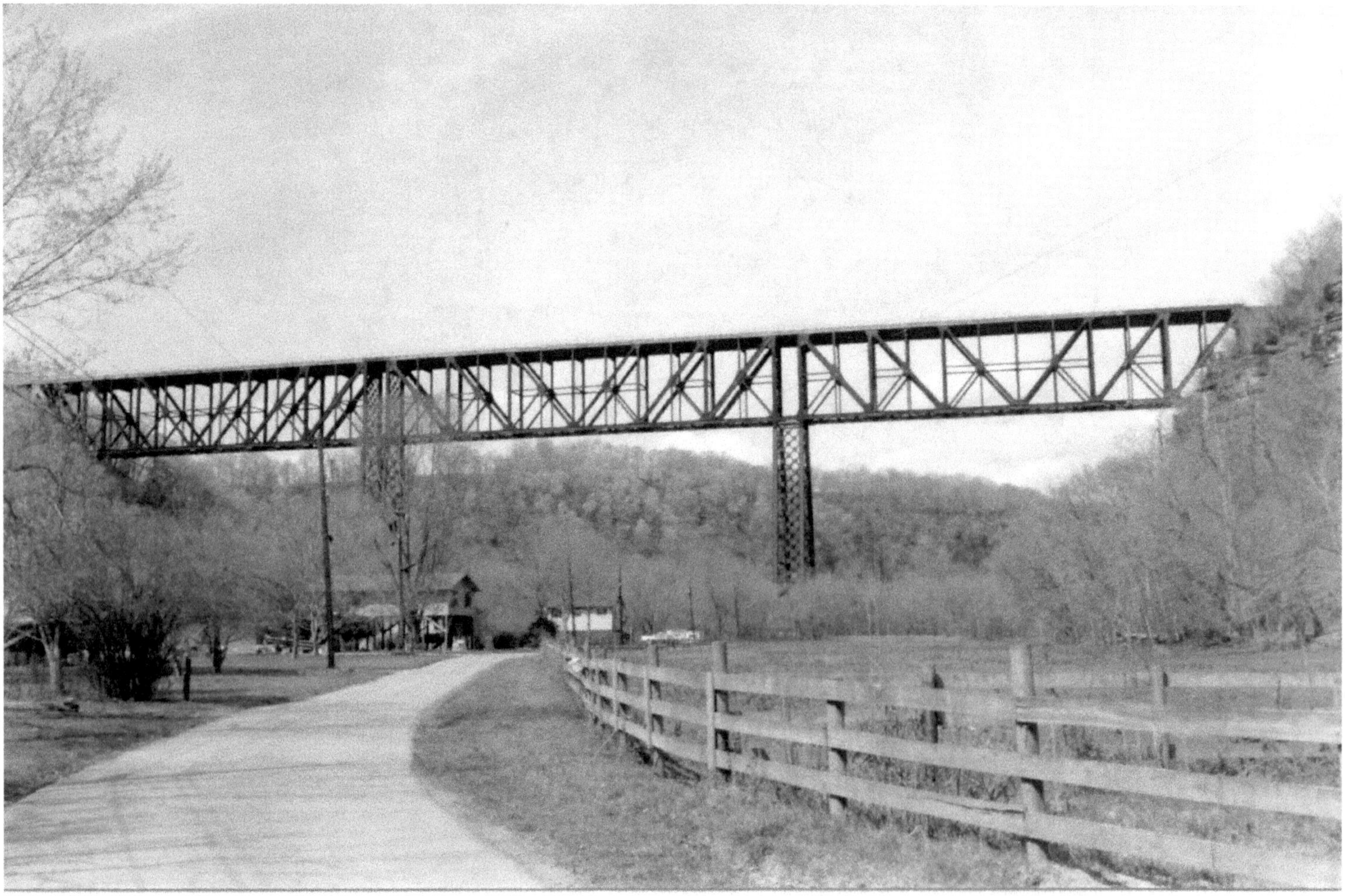

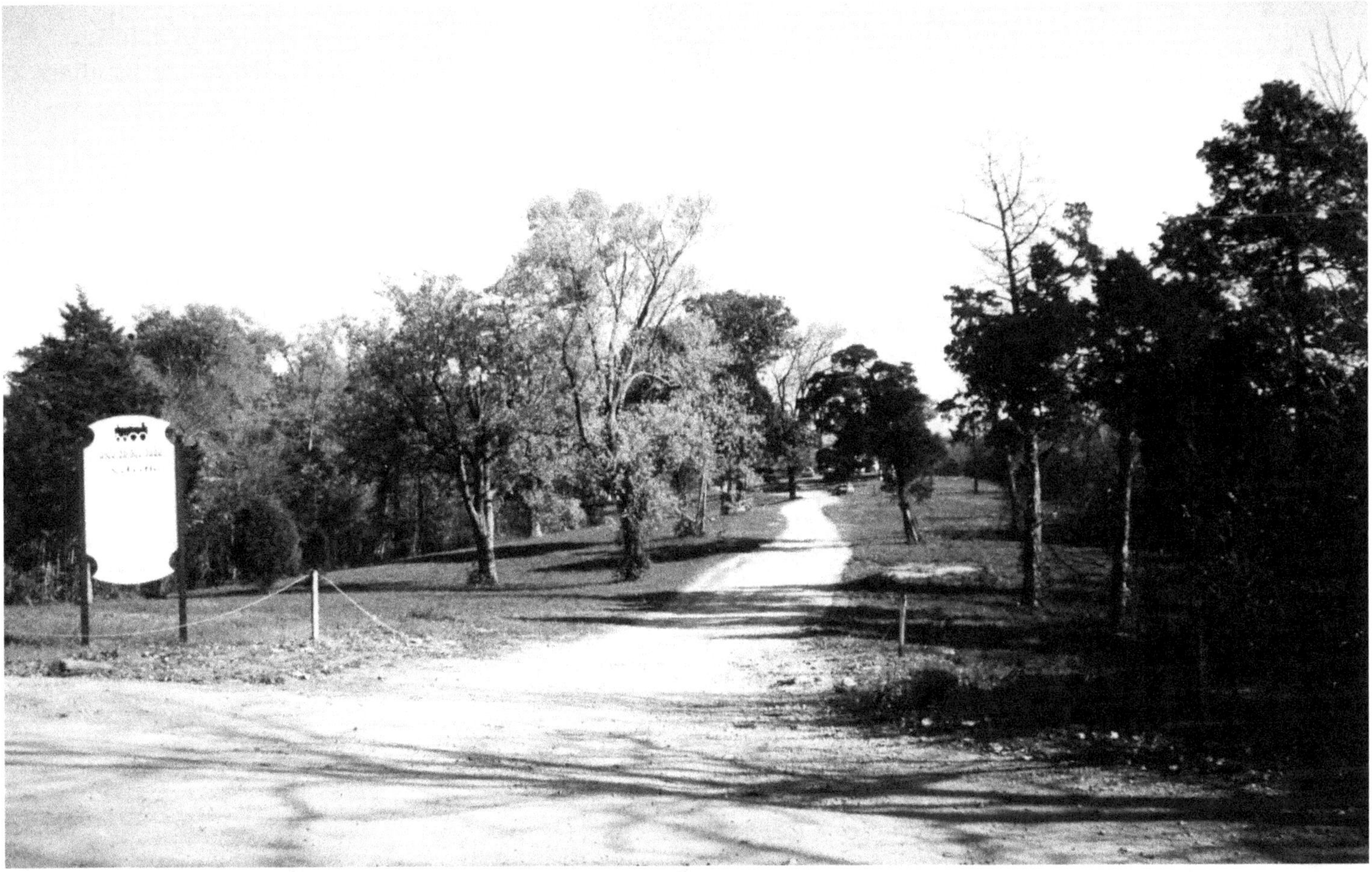

Beginning in the 1990's, interest in the park was renewed by local historians and politicians who wanted to preserve this treasure for generations to come.

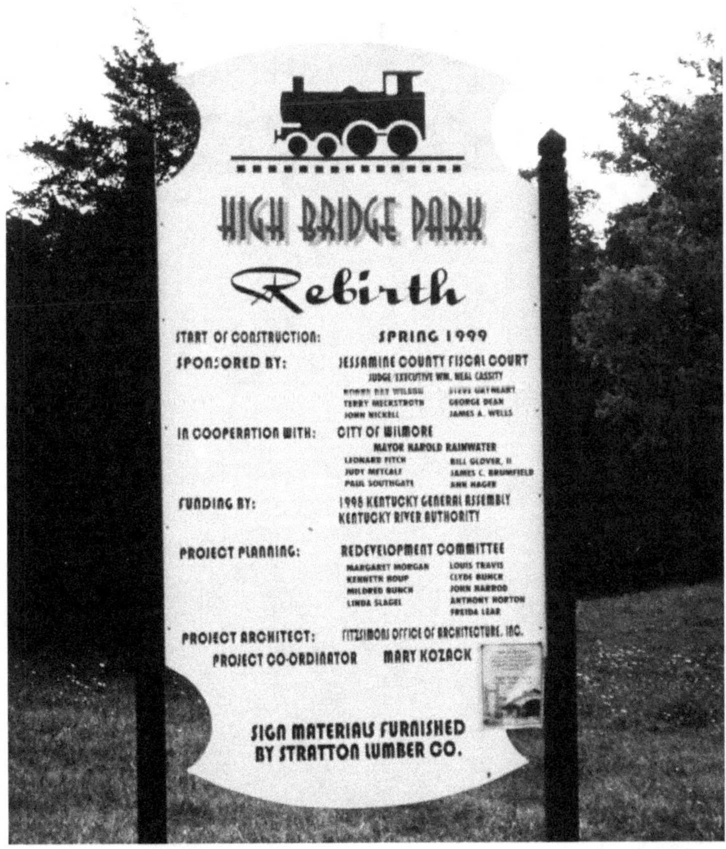

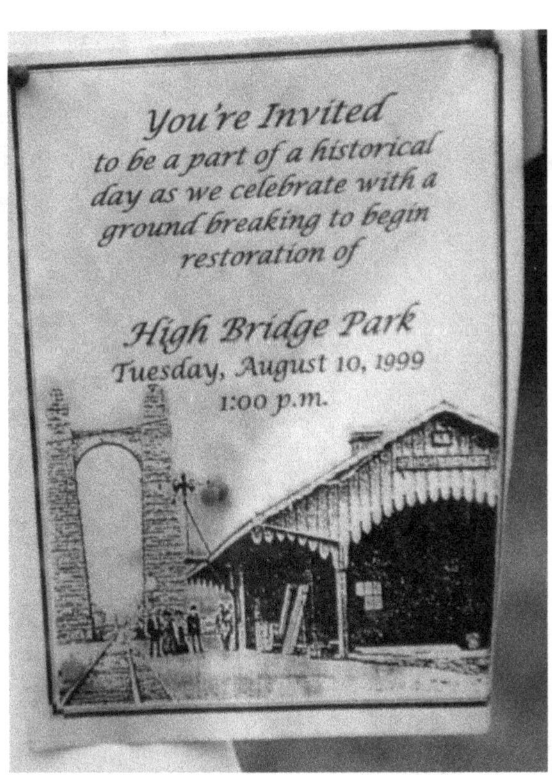

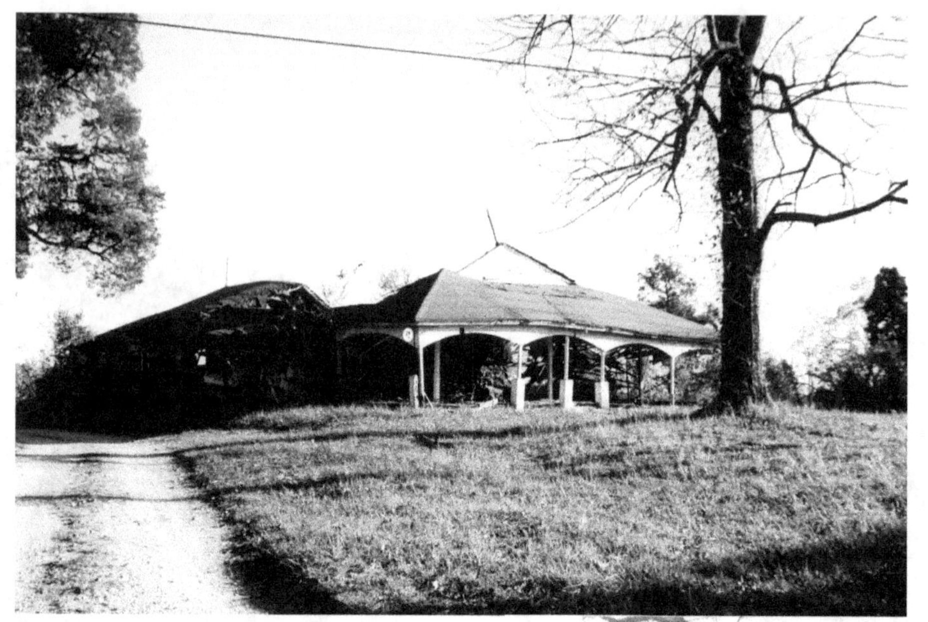

As part of the park's revitalization, the old, collapsed dance pavilion was razed and a new pavilion was built that is identical to the original but with some modern updates. The park and new pavilion were officially dedicated on Saturday, May 13, 2000, with hundreds in attendance, including this book's authors.

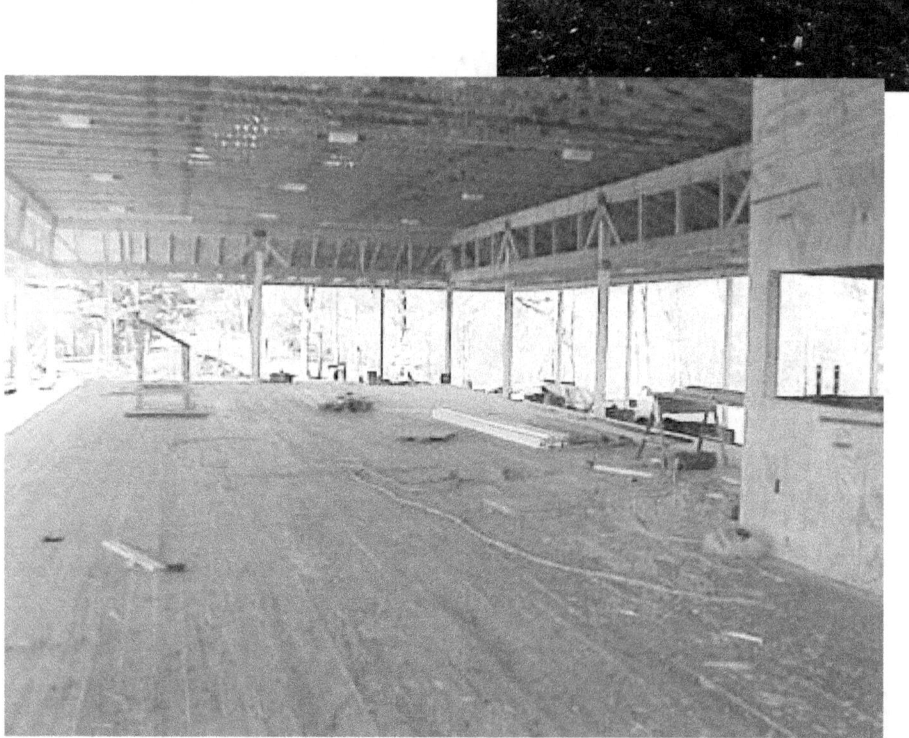

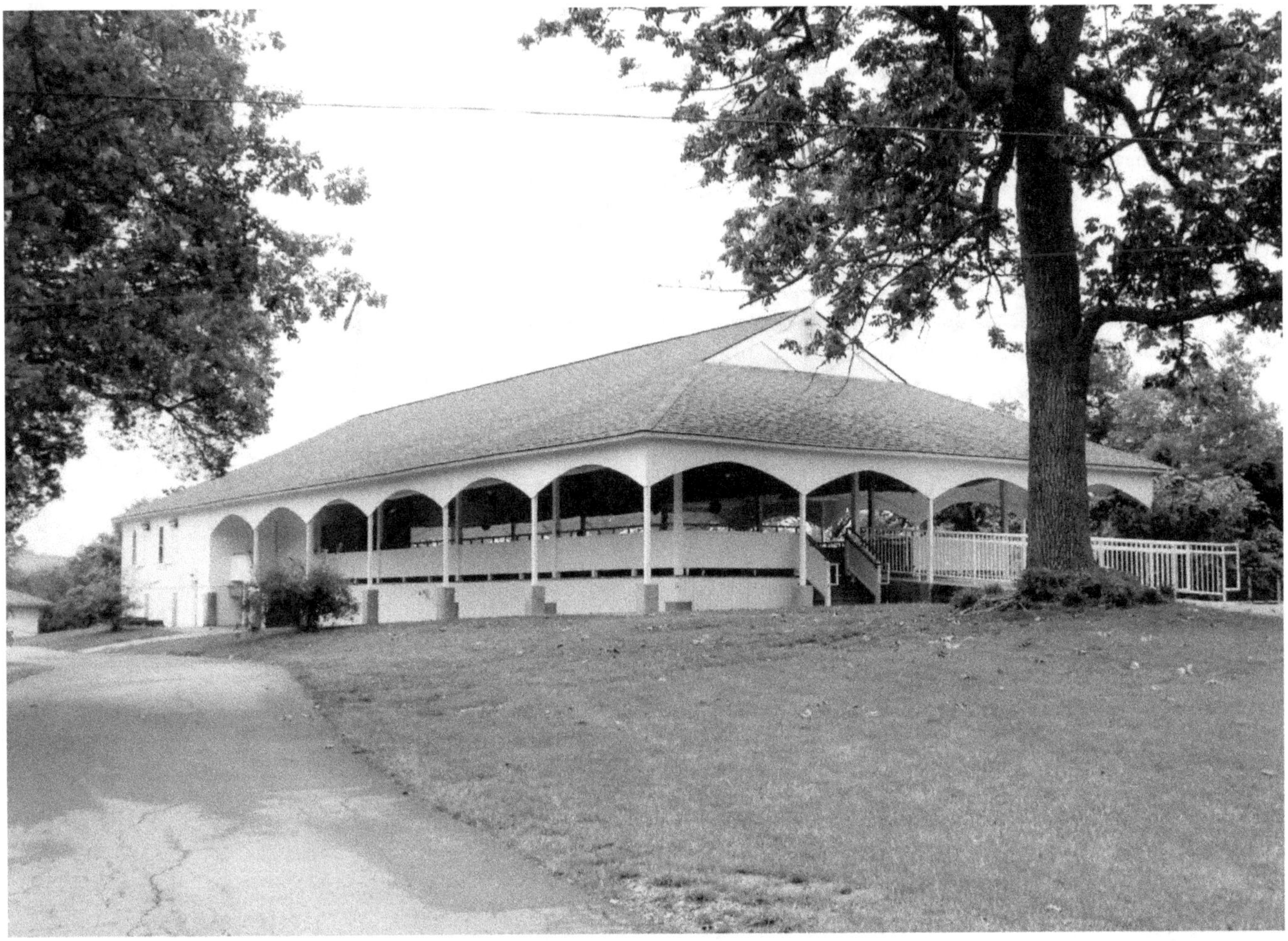

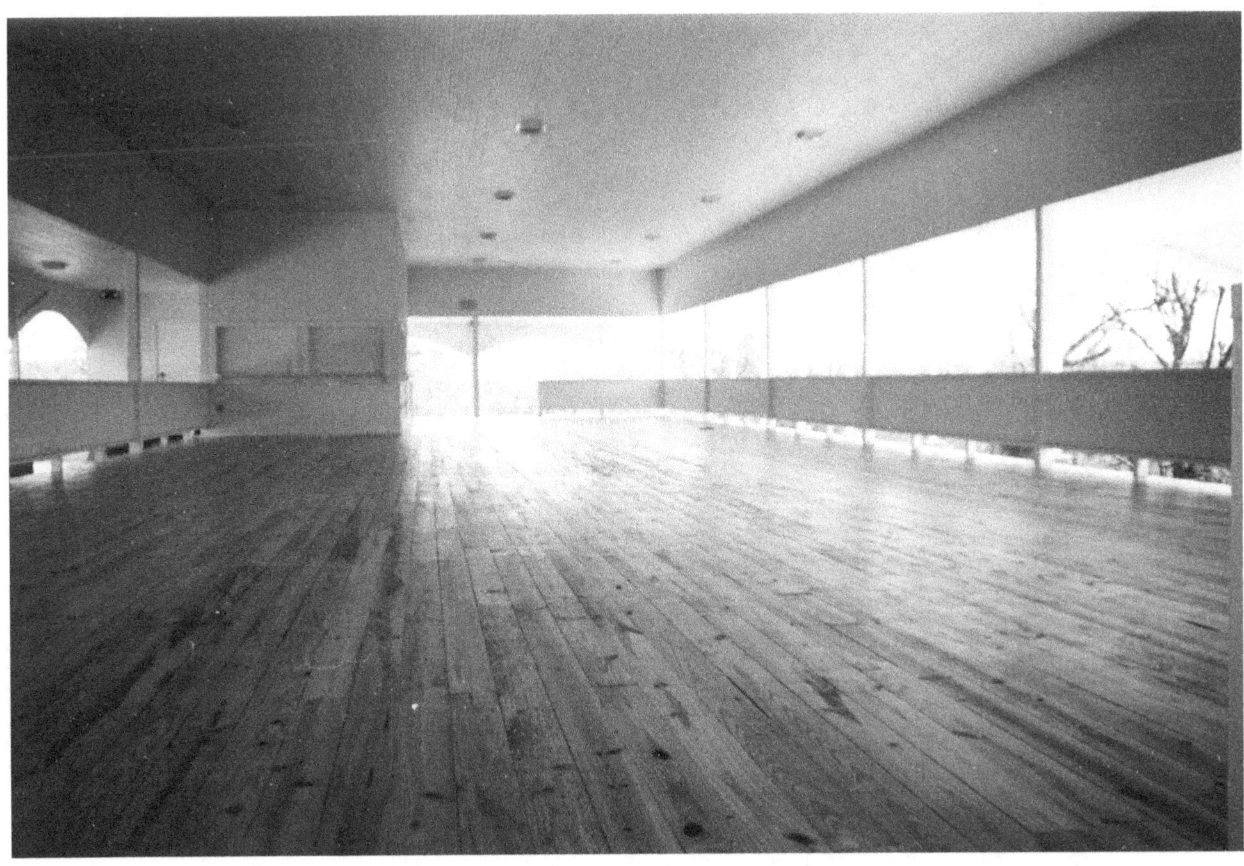

Program

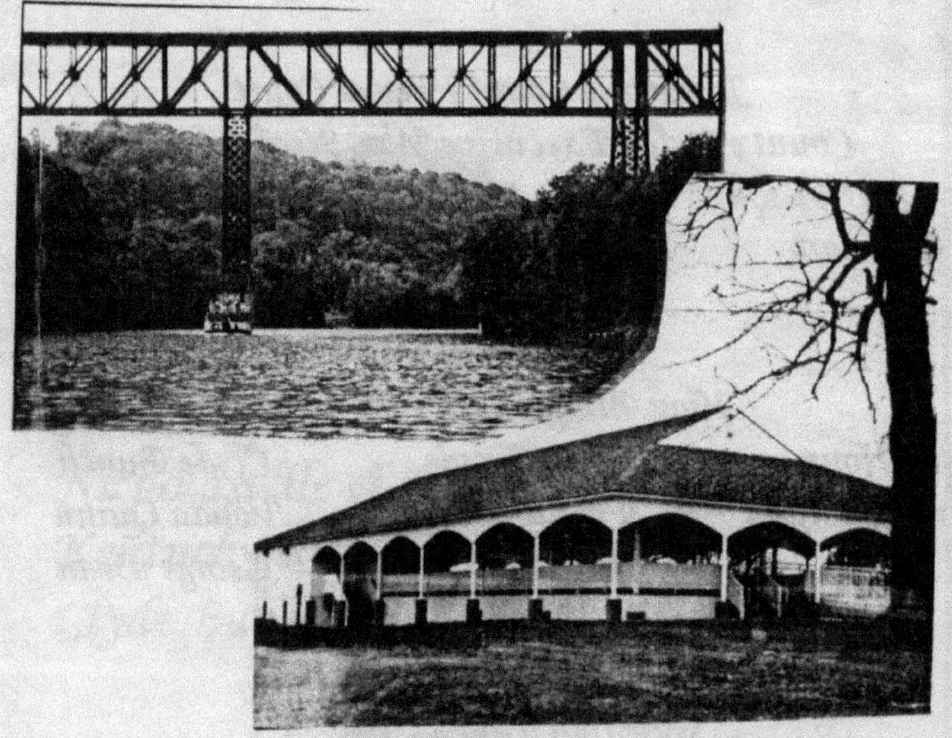

High Bridge Park Pavilion Dedication

Saturday May 13, 2000

The second phase of the rebirth of the park was repairing the neglected gazebos, adding paved parking areas, a scenic overlook and additional covered picnic shelters.

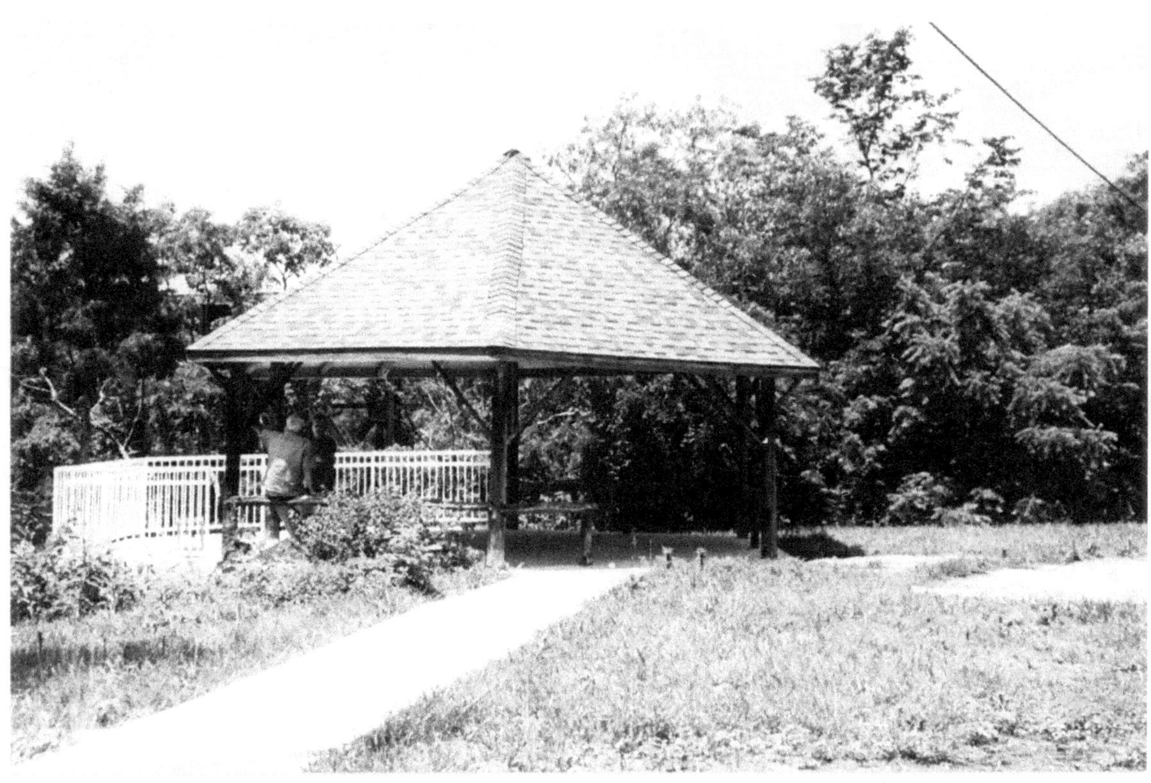

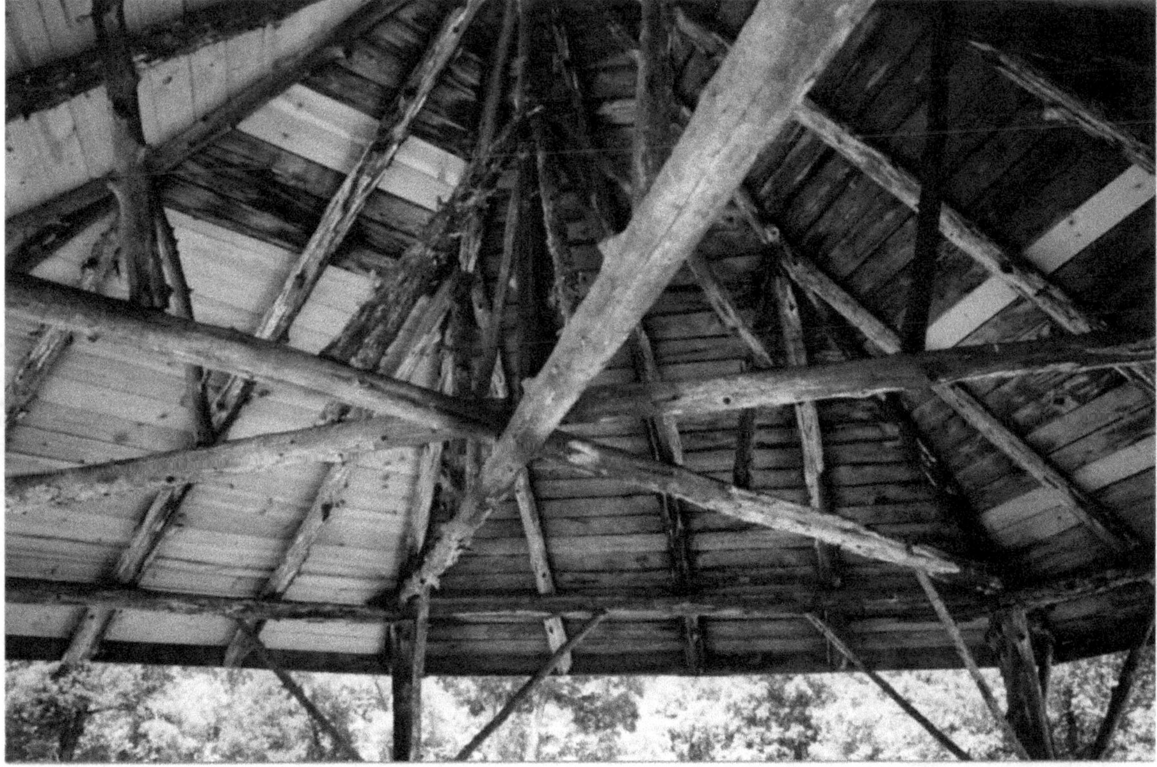

A combination of new and old materials were used in the renovation of the gazebos. This is an interior photo of one of their roofs.

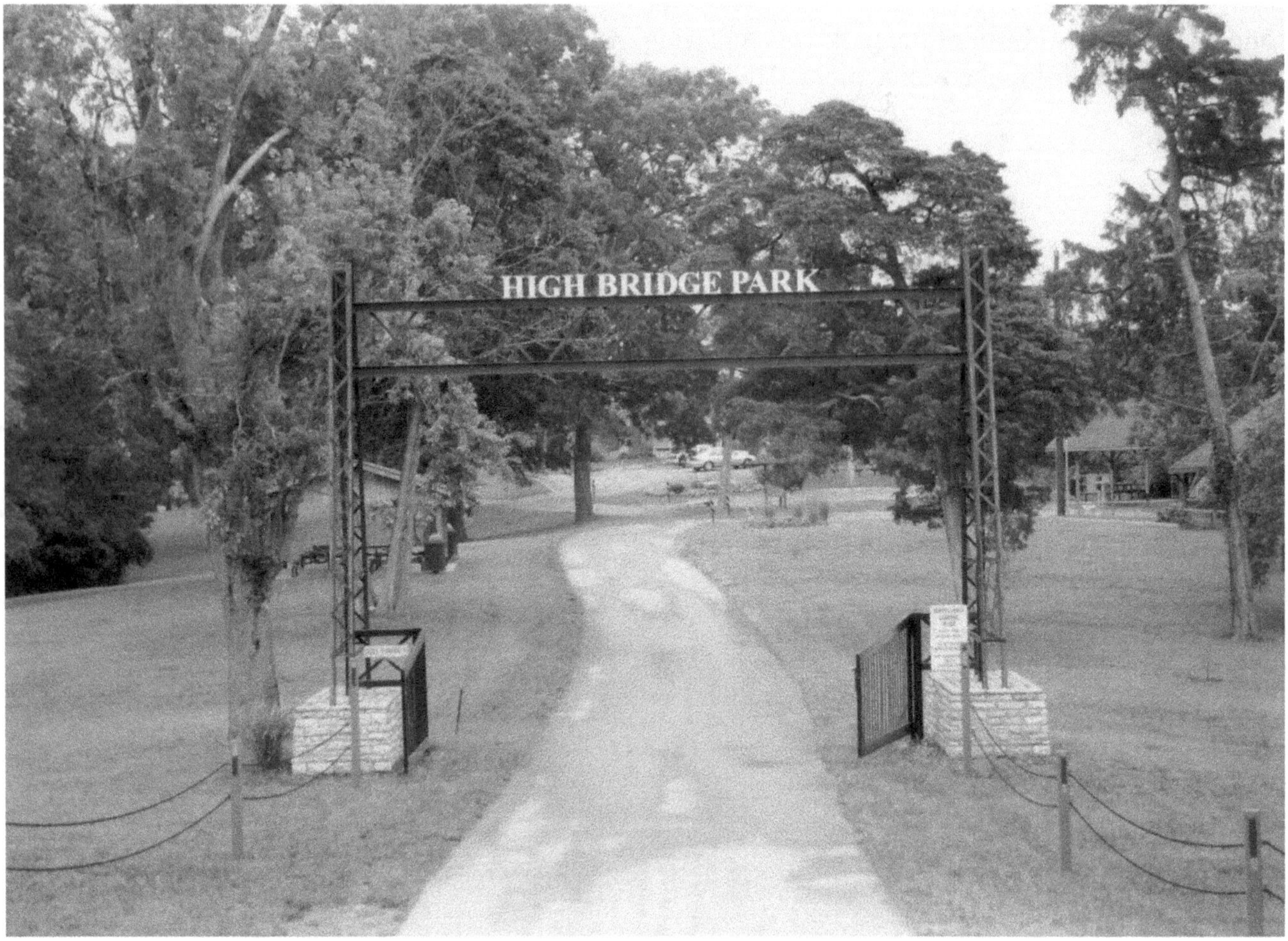

The park is open daily to the public and once again is a popular place for family reunions, wedding receptions, and picnics.

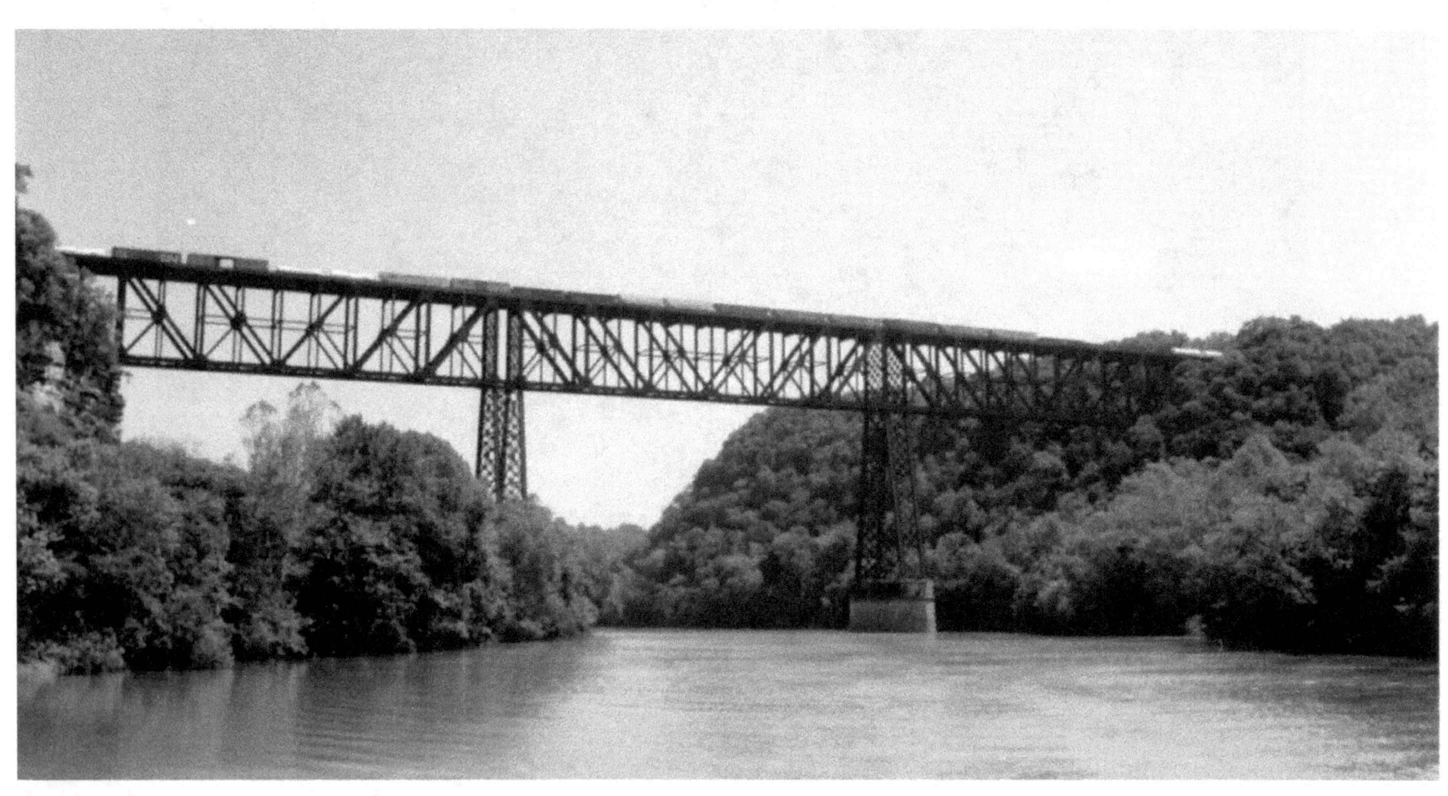

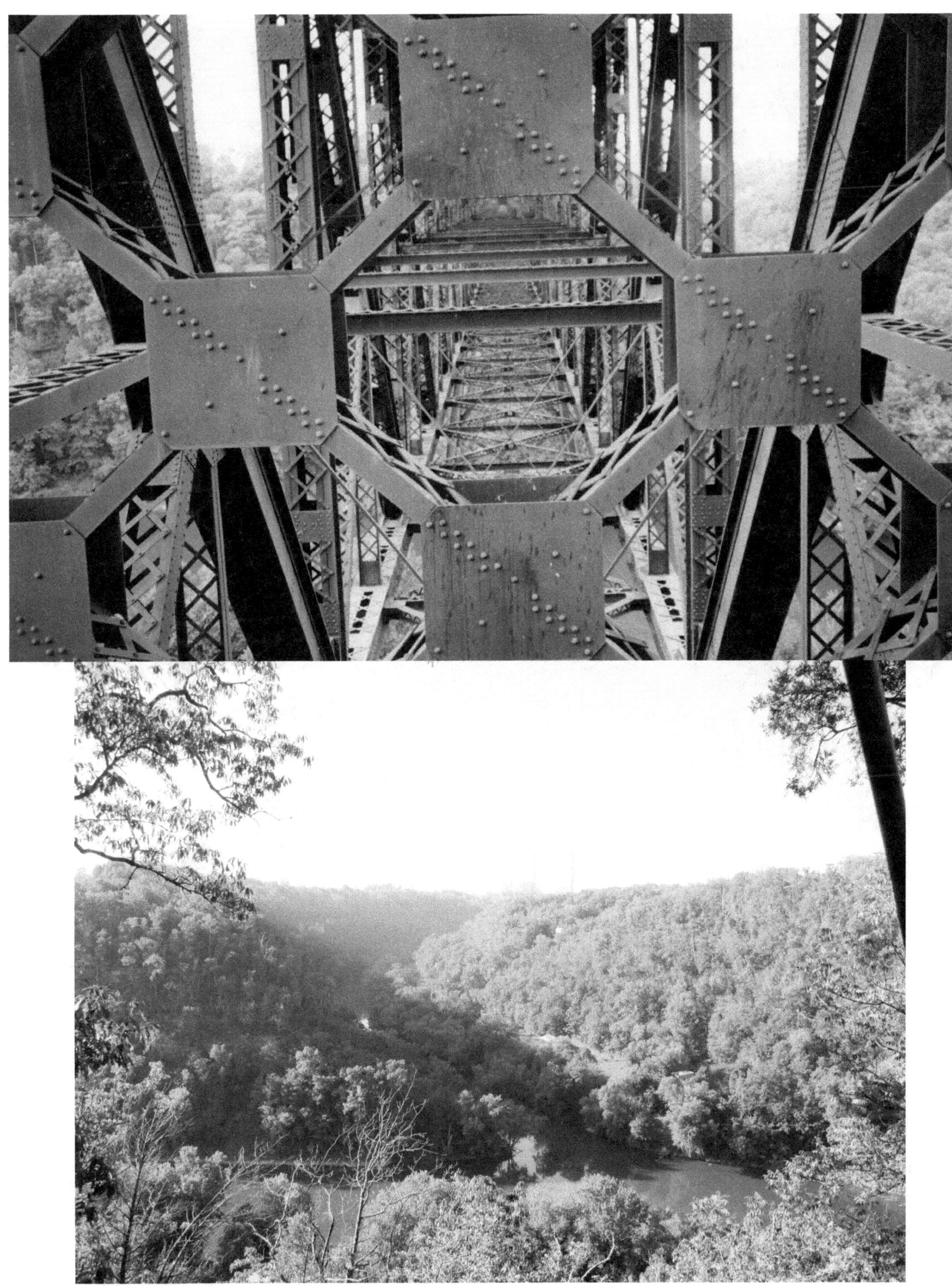

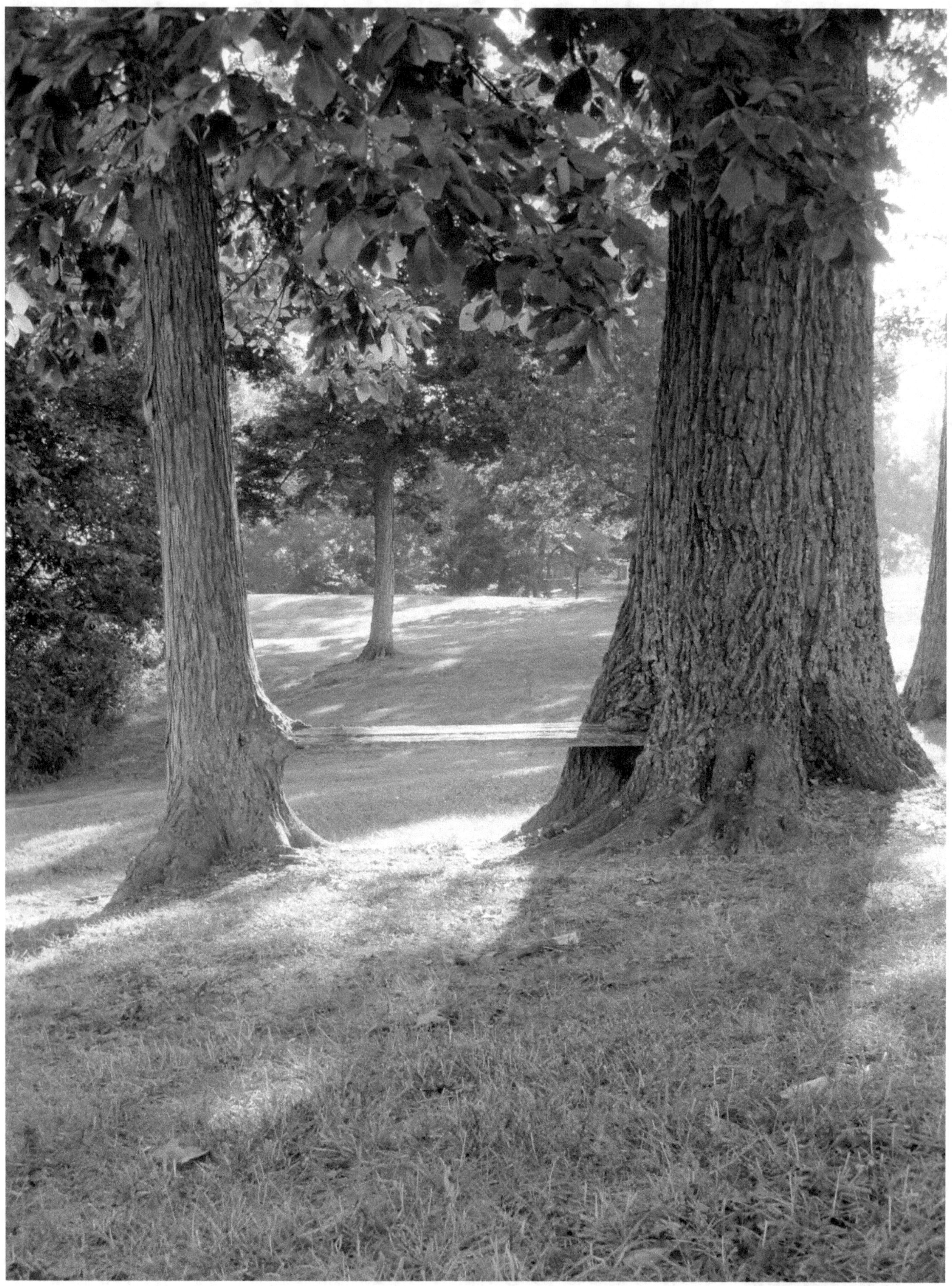

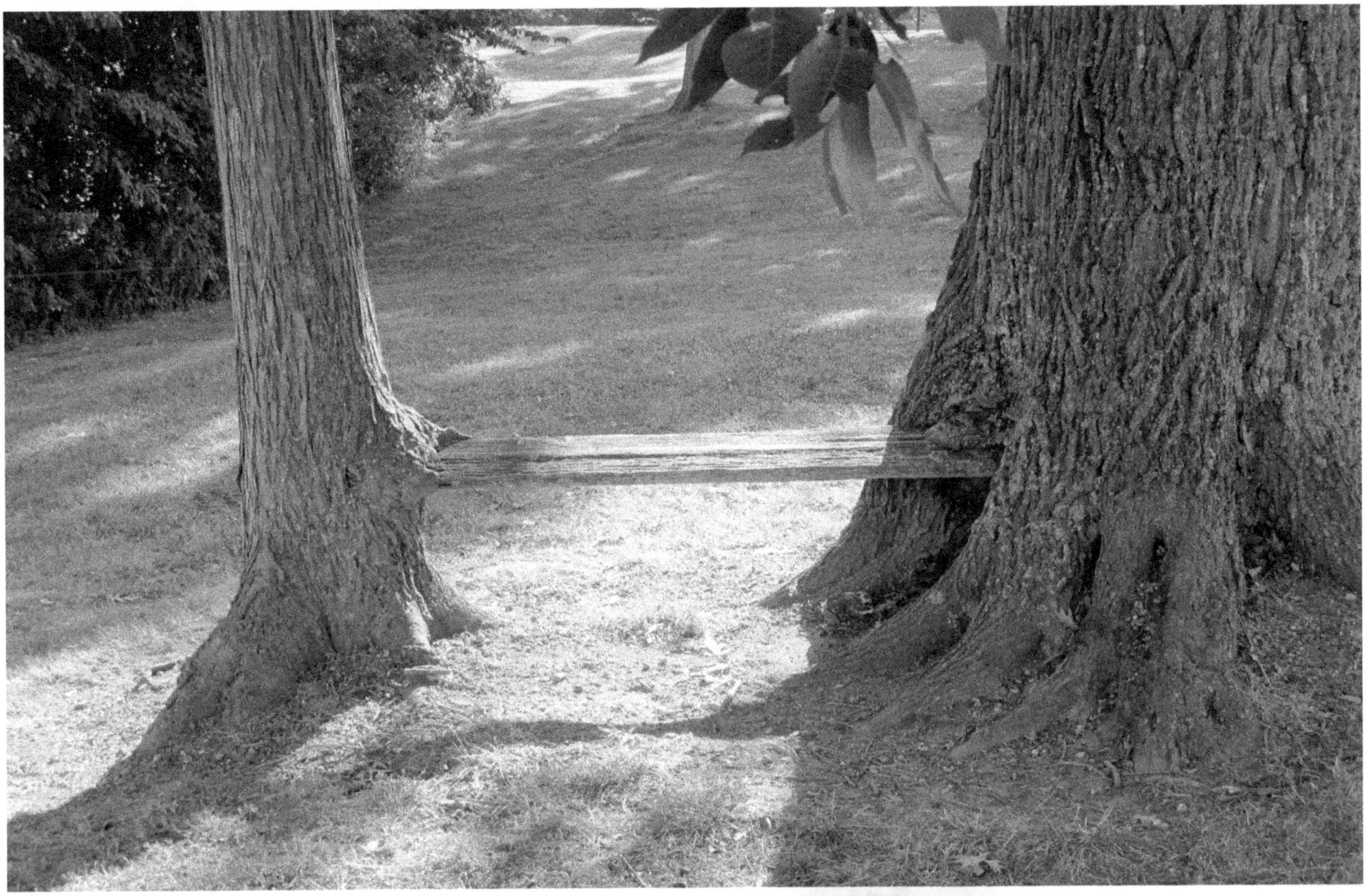

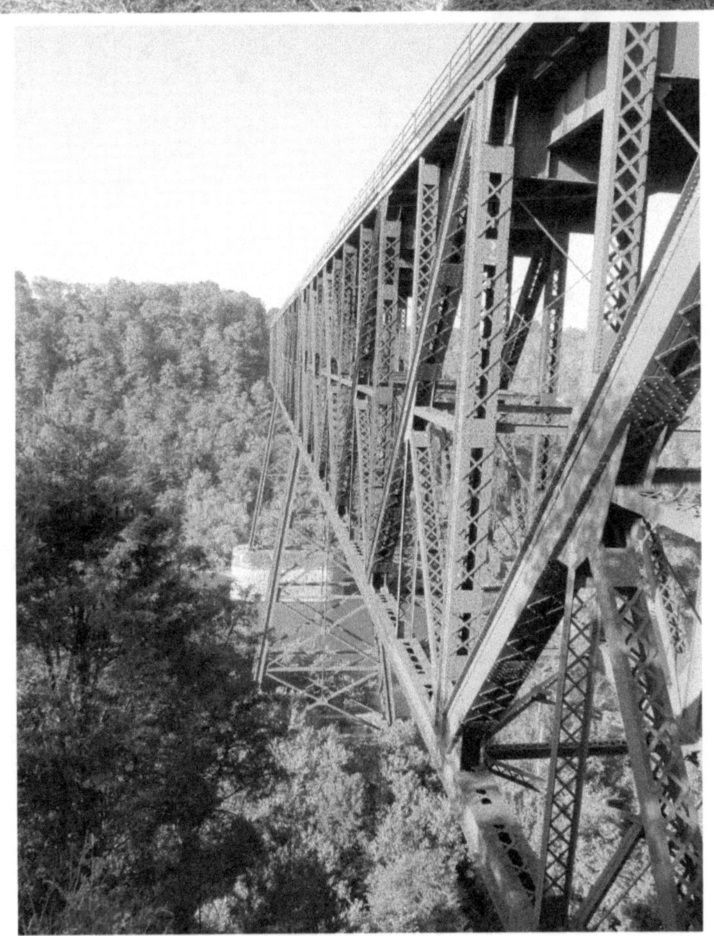

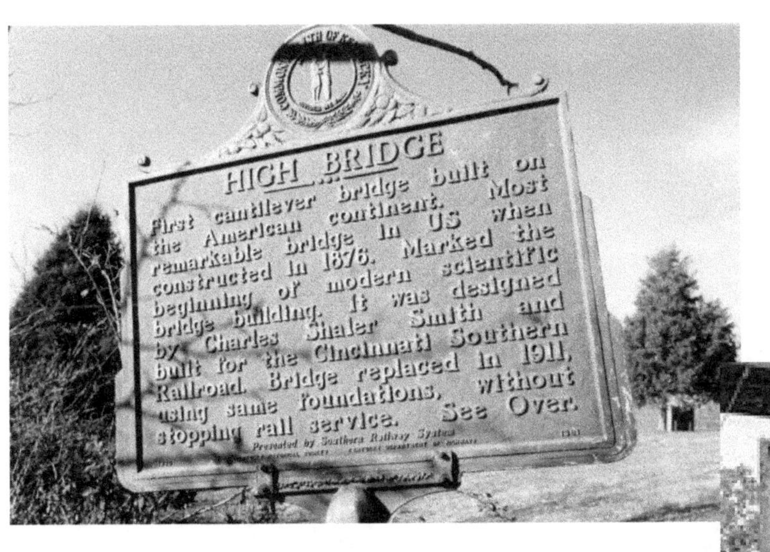
HIGH BRIDGE
First cantilever bridge built on the American continent. Most remarkable bridge in US when constructed in 1876. Marked the beginning of modern scientific bridge building. It was designed by Charles Shaler Smith and built for the Cincinnati Southern Railroad. Bridge replaced in 1911, using same foundations, without stopping rail service. See Over.
Presented by Southern Railway System

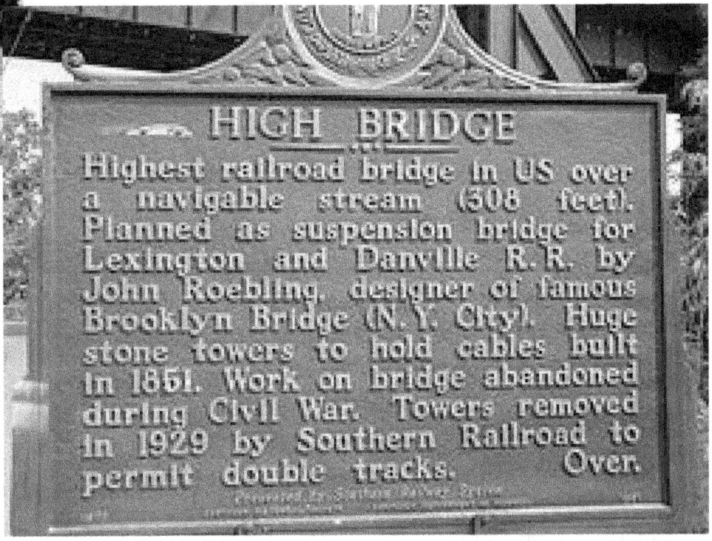

HIGH BRIDGE
Highest railroad bridge in US over a navigable stream (308 feet). Planned as suspension bridge for Lexington and Danville R.R. by John Roebling, designer of famous Brooklyn Bridge (N.Y. City). Huge stone towers to hold cables built in 1851. Work on bridge abandoned during Civil War. Towers removed in 1929 by Southern Railroad to permit double tracks. Over.

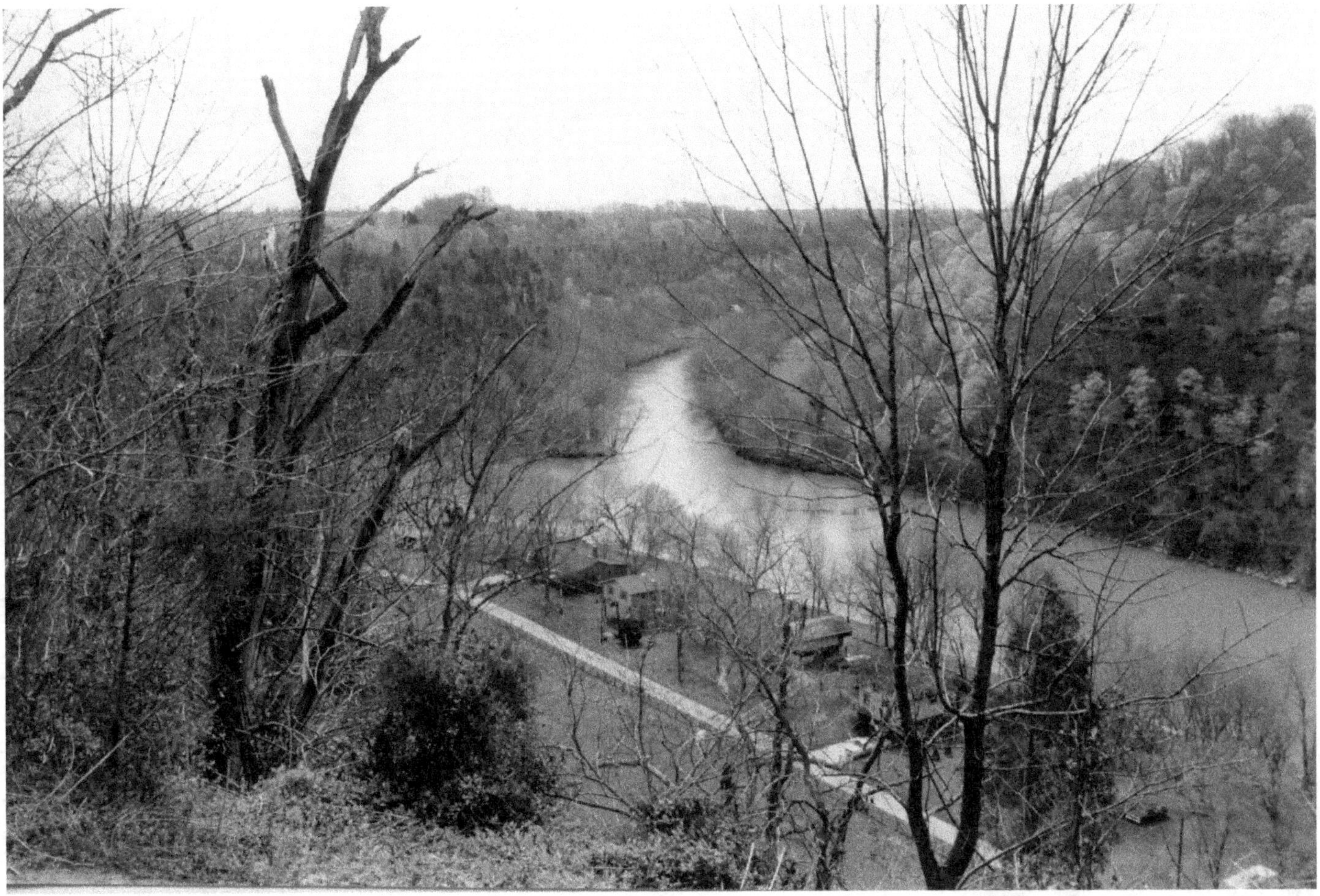

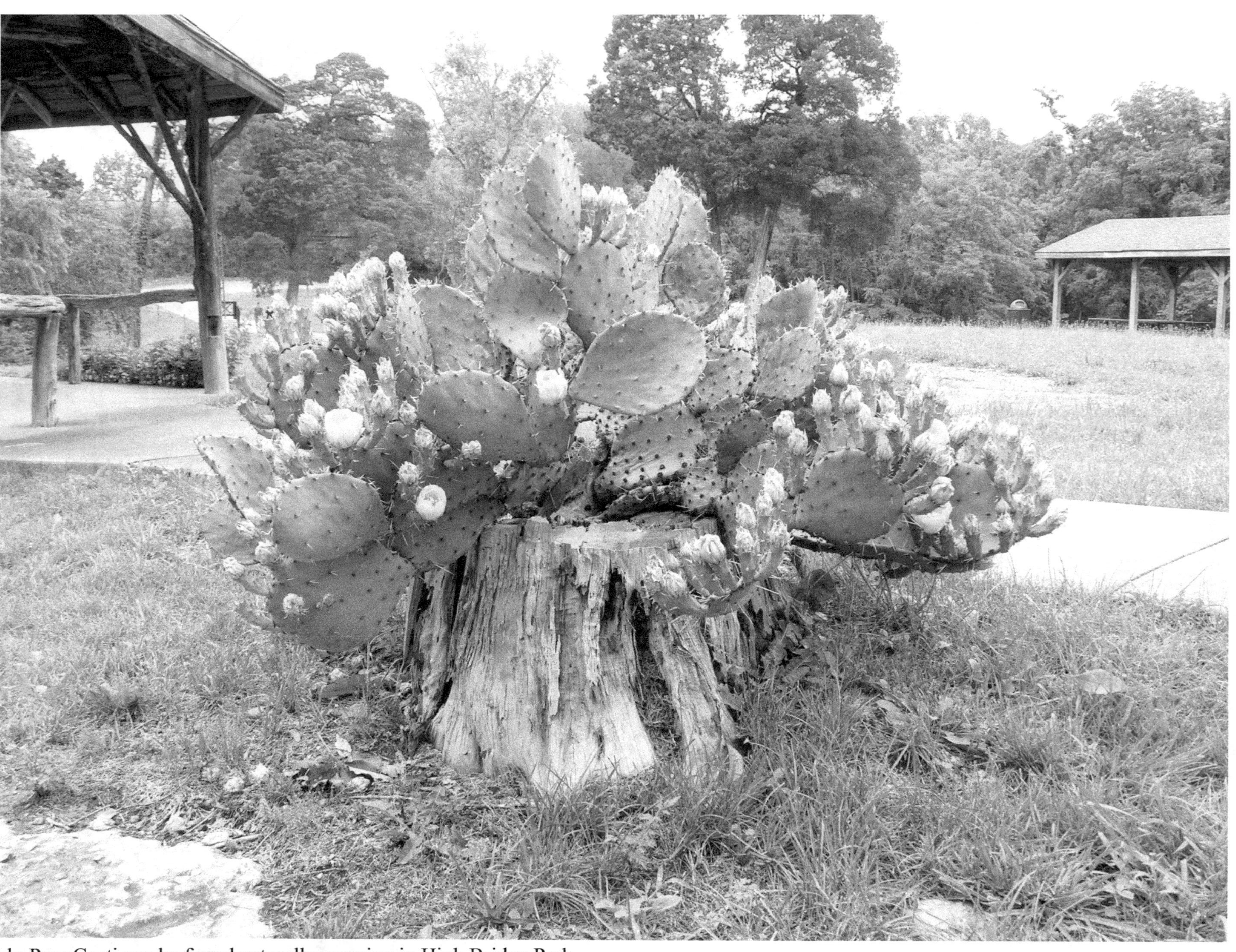

Prickly Pear Cacti can be found naturally growing in High Bridge Park.

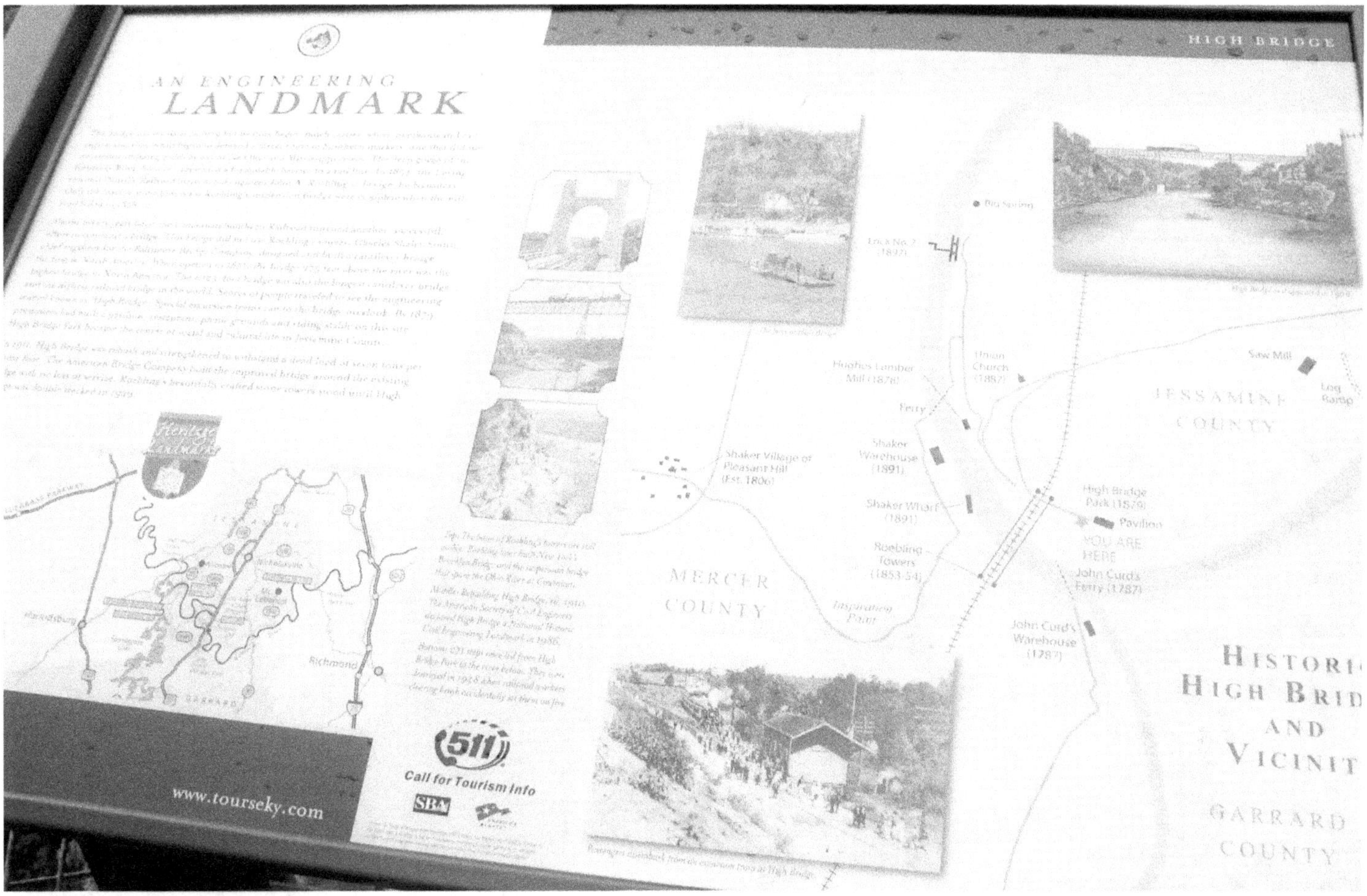

AN ENGINEERING LANDMARK

HIGH BRIDGE

MERCER COUNTY

JESSAMINE COUNTY

GARRARD COUNTY

www.tourseky.com

Call for Tourism Info

Information about the bridge, the park and the area can be found posted on the scenic overlook.

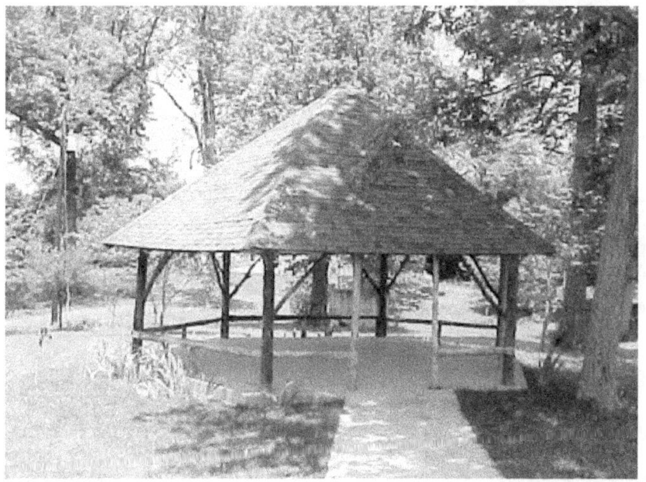

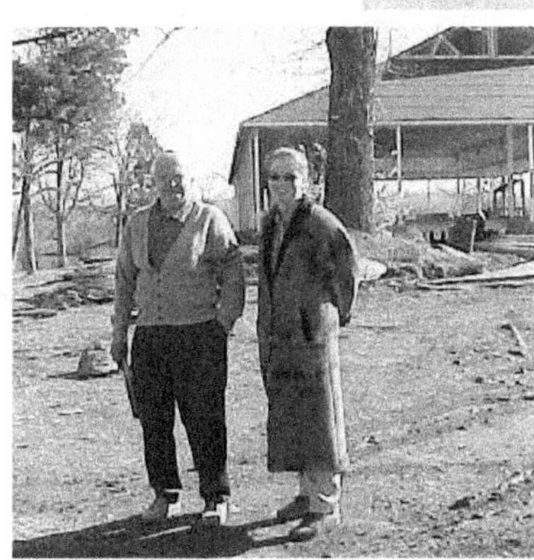

Clyde and Ken with the author.

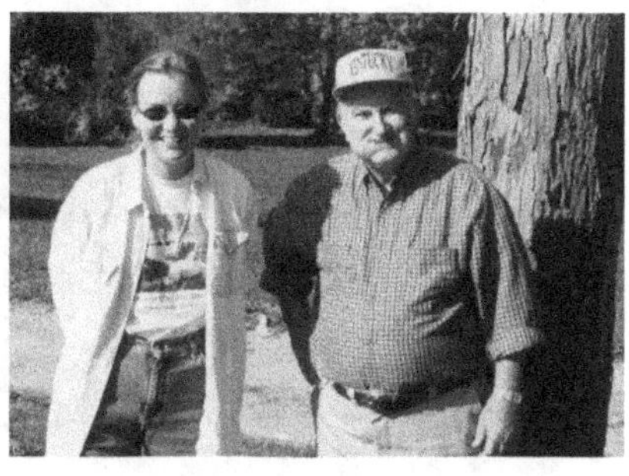

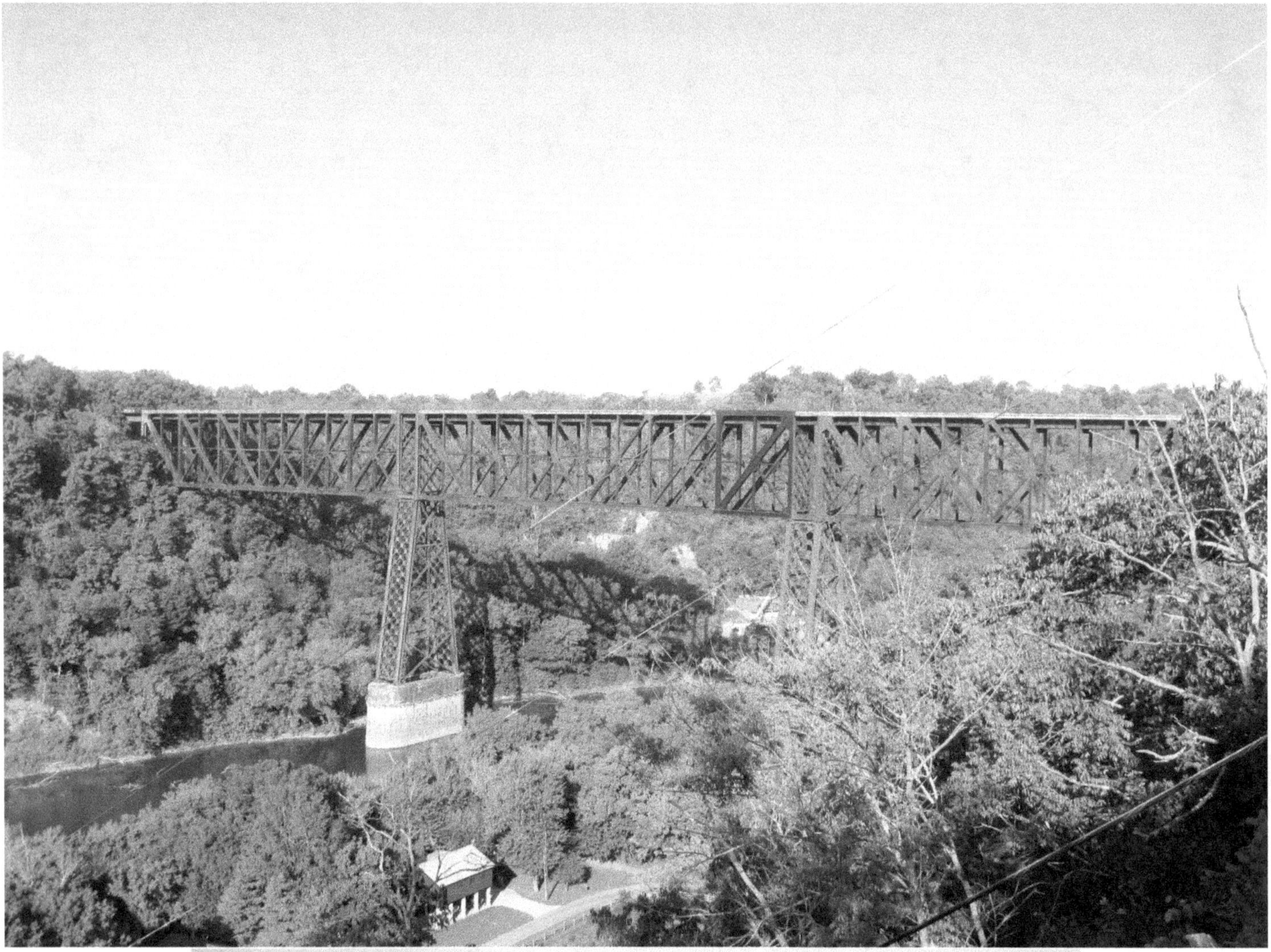

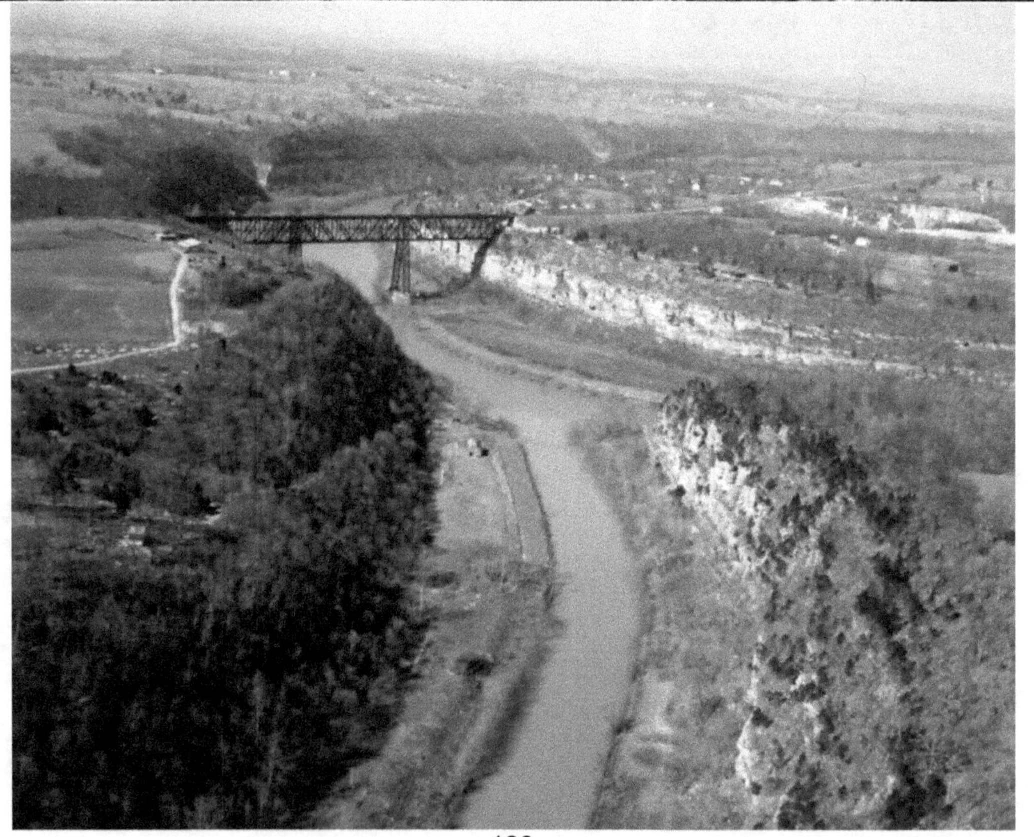

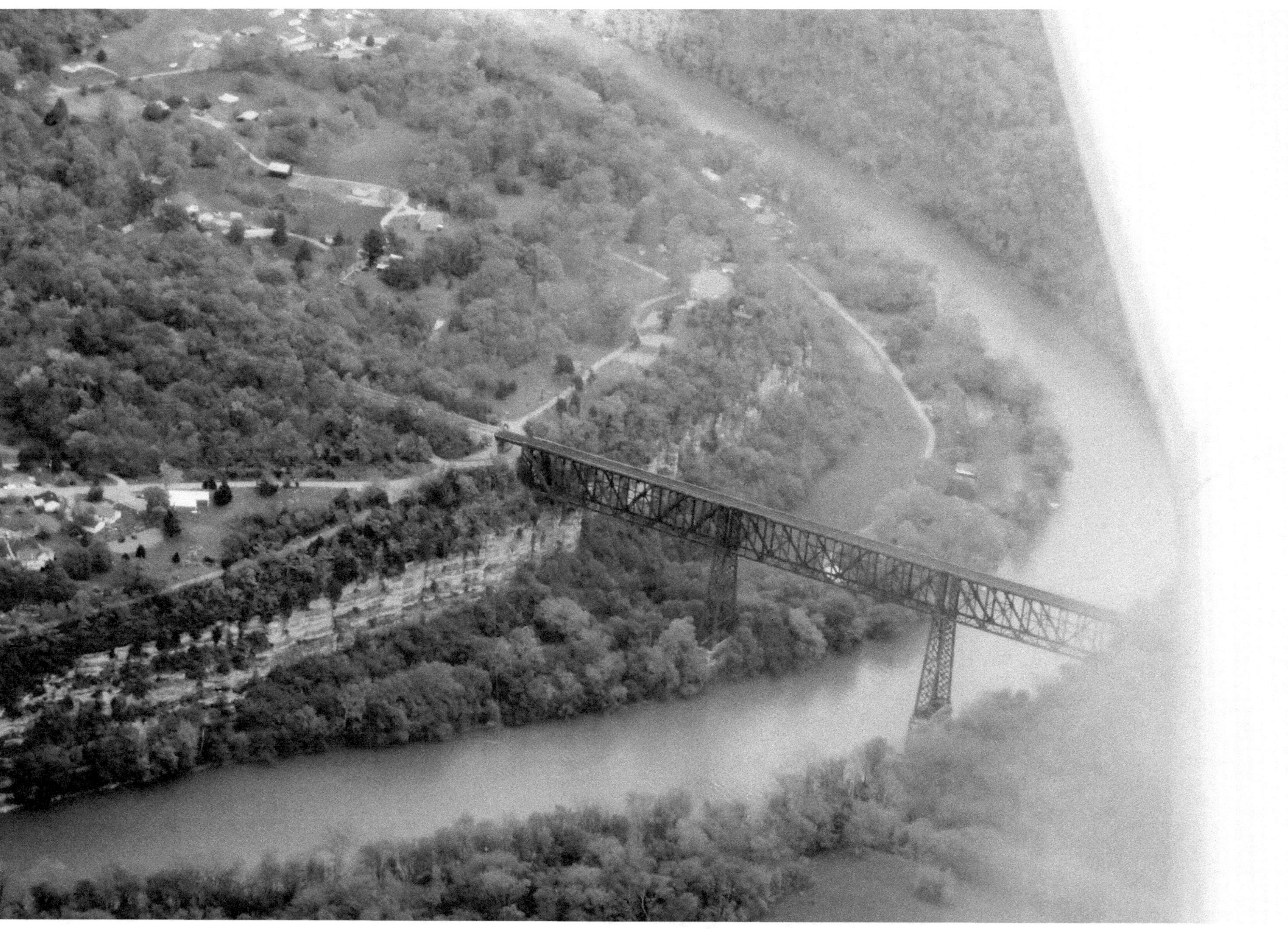

April 2012 view of High Bridge and the surrounding area taken on our commercial flight on its descent into Lexington, Kentucky.

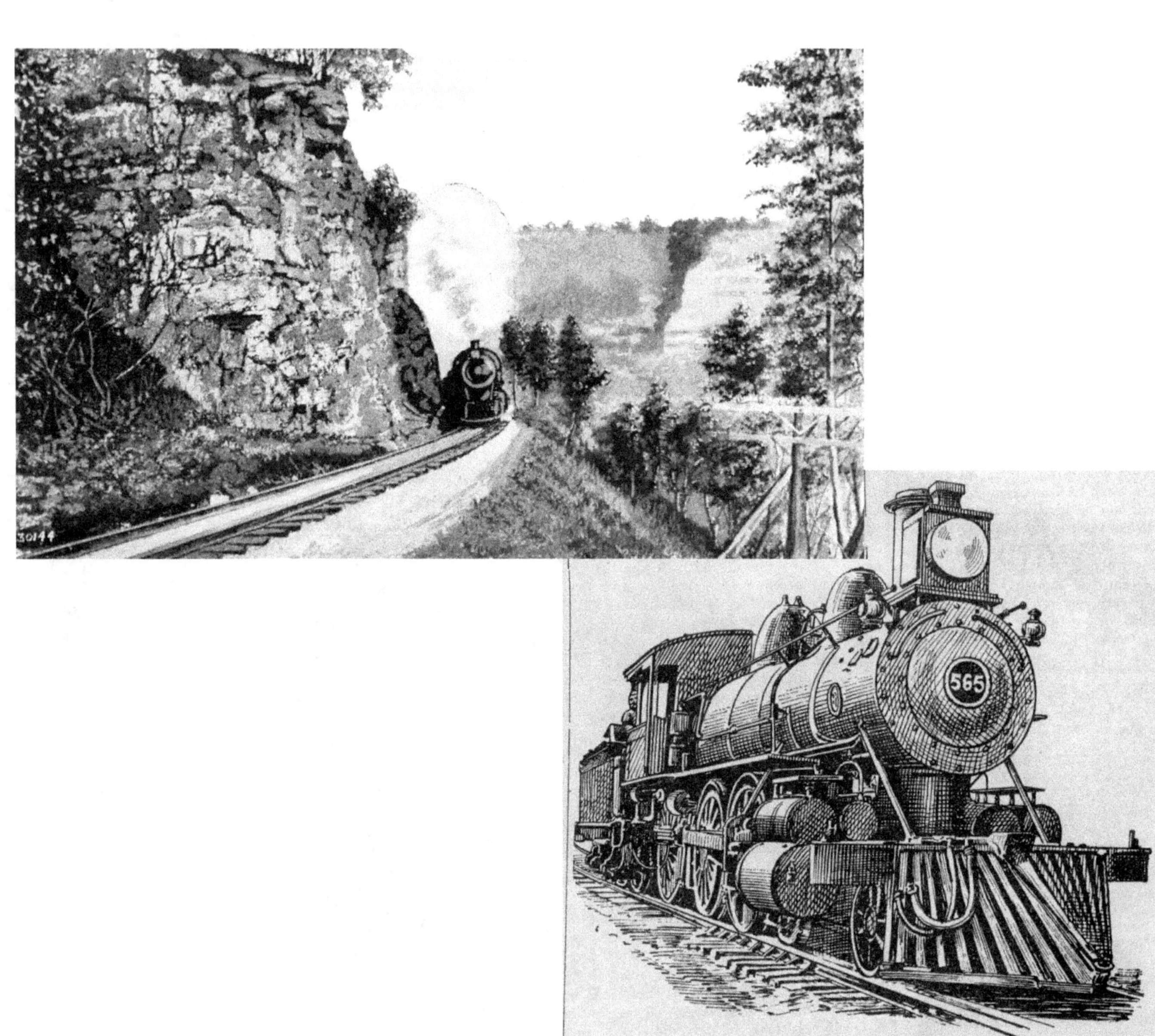

THE BIG MOGUL, QUEEN & CRESCENT.

About the Author

Melissa C. Jurgensen is the author of several books, including *River Towns of Central Kentucky* and *Covered Bridges of Fleming County, Kentucky*. Melissa was awarded the commission of Kentucky Colonel in 2006, the highest honor in the Commonwealth, due to her advocacy for the restoration and preservation of Kentucky's covered bridges. Melissa also serves on the board of the Harrodsburg Historical Society, and as an Executive Bourbon Steward, she has a keen interest in the history and evolution of Kentucky's all important bourbon industry.

Photo Credits

Melissa C. Jurgensen, Judith G. Jurgensen, Kenneth Houp, Clyde Bunch, University of Kentucky Photographic Archives, William Coffey (Paul Sawyier Art Galleries), Jerry Sampson and Amalie Preston.

www.ingramcontent.com/pod-product-compliance
Lightning Source LLC
LaVergne TN
LVHW061249100826
845148LV00008B/1068